T0186826

COASTAL AQUIFER MANAGEMENT

monitoring, modeling, and case studies

COASTAL AQUIFER MANAGEMENT

monitoring, modeling, and case studies

Edited by
Alexander H.-D. Cheng
Driss Ouazar

LEWIS PUBLISHERS

A CRC Press Company
Boca Raton   London   New York   Washington, D.C.

Library of Congress Cataloging-in-Publication Data

Coastal aquifer management ; monitoring, modeling, and case studies / edited by
Alexander H.-D. Cheng, Driss Ouazar.
    p. cm.
Includes bibliographical references and index.
ISBN 1-56670-605-X
    1. Aquifers. 2. Saltwater encroachment. 3. Groundwater flow. 4. Coastal zone
management. I. Cheng, A. H.-D. II. Ouazar, D.

GB1199.C63 2003
363.739'4—dc22   2003060544

This book contains information obtained from authentic and highly regarded sources. Reprinted material is quoted with permission, and sources are indicated. A wide variety of references are listed. Reasonable efforts have been made to publish reliable data and information, but the author and the publisher cannot assume responsibility for the validity of all materials or for the consequences of their use.

Neither this book nor any part may be reproduced or transmitted in any form or by any means, electronic or mechanical, including photocopying, microfilming, and recording, or by any information storage or retrieval system, without prior permission in writing from the publisher.

All rights reserved. Authorization to photocopy items for internal or personal use, or the personal or internal use of specific clients, may be granted by CRC Press LLC, provided that $1.50 per page photocopied is paid directly to Copyright Clearance Center, 222 Rosewood Drive, Danvers, MA 01923 USA. The fee code for users of the Transactional Reporting Service is ISBN 1-56670-605-X/03/$0.00+$1.50. The fee is subject to change without notice. For organizations that have been granted a photocopy license by the CCC, a separate system of payment has been arranged.

The consent of CRC Press LLC does not extend to copying for general distribution, for promotion, for creating new works, or for resale. Specific permission must be obtained in writing from CRC Press LLC for such copying.

Direct all inquiries to CRC Press LLC, 2000 N.W. Corporate Blvd., Boca Raton, Florida 33431.

Trademark Notice: Product or corporate names may be trademarks or registered trademarks, and are used only for identification and explanation, without intent to infringe.

Visit the CRC Press Web site at www.crcpress.com

© 2004 by CRC Press LLC
Lewis Publishers is an imprint of CRC Press LLC

No claim to original U.S. Government works
International Standard Book Number 1-56670-605-X
Library of Congress Card Number 2003060544
Printed in the United States of America 1 2 3 4 5 6 7 8 9 0
Printed on acid-free paper

Preface

About 70% of the world's population dwells in coastal zones. With the economic and population growth, the shortage in freshwater supply becomes increasingly acute. With surface water more and more depleted and polluted, coastal communities have turned to groundwater to make up for the shortfall. For domestic supply purposes, the percentage of groundwater use has increased to more than 40% on a worldwide basis.

Coastal aquifers are highly sensitive to anthropogenic disturbances. Inappropriate management of coastal aquifers can lead to irreversible damages, leading to their destruction as freshwater sources. Being aware of the threat, federal, state, and local water agencies have intensified saltwater intrusion monitoring and prevention projects, and increased coastal aquifer planning and management efforts. In the last two decades, a significant amount of knowledge has been accumulated and new technologies were developed. This book is an effort to assemble these advancements in order to share them with the communities, the technical profession, and the water supply industry, as well as governmental regulators and policy makers.

This book may be viewed as a sequel to the first book published on this subject: *Seawater Intrusion in Coastal Aquifers—Concepts, Methods and Practices*, by Bear, Cheng, Sorek, Ouazar, and Herrera (Kluwer, 1999). The first book presented the basic concepts, theories, and methodologies, which can be used as a textbook for learning this subject. The current book focuses on practical experiences.

In the year 2001, the *First International Conference on Saltwater Intrusion and Coastal Aquifers—Monitoring, Modeling and Management* was convened at the quiet, historical coastal town of Essaouira, in Morocco. From the participants of the conference, a group of international experts who were practitioners in federal and state water agencies, consulting companies, research laboratories, and universities was assembled to contribute to the book. This international panel was further joined by a few participants of the second conference held in Merida, Mexico, in 2003.

The 12 chapters collected cover a broad spectrum, ranging from hydrogeology, geochemistry, geophysics, optimization, uncertainty analysis, GIS, monitoring, and computer modeling, to planning and management. Each chapter is based on case studies that provide worldwide experiences from practices in the Gaza Strip, Italy, Spain, the Netherlands, Mexico, and U.S. communities in California, Florida, Massachusetts, and Alaska.

In addition to the above, this book contains another innovation. It is among the first to be co-published with a CD. Due to the large amount of data, color graphics, computer programs, documentation, and other materials associated with the field studies reported in the book, a conventionally printed book is no longer adequate. With the CD, the capacity of the book can be augmented. Not only can multimedia materials be presented in color, in animation, etc., these materials can also be easily updated in the future. We hope that through this combination of traditional and modern presentation techniques, we can bring the best of both to the reader.

Alexander H.-D. Cheng
Oxford, Mississippi, USA

Driss Ouazar
Rabat, Morocco

August, 2003

Contributing Authors

Mark Bakker
University of Georgia, USA

Mark Barcelo
Southwest Florida Water Management District, USA

Giovanni Barrocu
University of Cagliari, Italy

David Andrew Barry
University of Edinburgh, UK

Michael Beach
Southwest Florida Water Management District, USA

Mohammed Karim Benhachmi
Ecole Mohammadia d'Ingénieurs, Morocco

Jenny Chapman
Desert Research Institute, Nevada, USA

Alexander H.-D. Cheng
University of Mississippi, USA

Khalid EL Harrouni
Ecole Nationale d'Architecture, LabHAUT, Morocco

Hedeff I. Essaid
U.S. Geological Survey, Menlo Park, California, USA

Robert Fitzgerald
Camp Dresser & McKee, Inc., USA

Driss Halhal
Water and Electricity Distribution Co., Tangier, Morocco

Brendan Harley
Camp Dresser & McKee, Inc., USA

Ahmed Hassan
Desert Research Institute, Nevada, USA

Brian Heywood
Camp Dresser & McKee, Inc., USA

Rehad Hossain
Camp Dresser & McKee, Inc., USA

Dong-Sheng Jeng
Griffith University, Australia

Theodore A. Johnson
Water Replenishment of Southern California, USA

Walter Jones
HydroGeoLogic Inc., USA

Jack L. Kindinger
U.S. Geological Survey, St. Petersburg, Florida, USA

Christian D. Langevin
U.S. Geological Survey, Miami, Florida, USA

Ling Li
University of Queensland, Brisbane, Australia

Mark Maimone
Camp Dresser & McKee, Inc., USA

Luis E. Marin
Universidad Nacional Autónoma de México, México

Henning Moe
Camp Dresser & McKee, Inc., USA

Laura Muscas
Center for Advanced Studies, Research and Development in Sardinia, Cagliari, Italy

Ahmed Naji
Faculté des Sciences et Techniques de Tanger, Morocco

Driss Ouazar
Ecole Mohammadia d'Ingénieurs, Morocco

Gualbert H.P. Oude Essink
Netherlands Institute of Applied Geosciences
Free University of Amsterdam, The Netherlands

Frederick L. Paillet
University of Maine, USA

Sorab Panday
HydroGeoLogic Inc., USA

Eugene C. Perry
Northern Illinois University, USA

Karl Pohlmann
Desert Research Institute, Nevada, USA

Henning Prommer
University of Edinburgh, UK

Maria Grazia Sciabica
University of Cagliari, Italy

Birgit Steinich
Universidad Nacional Autónoma de México, México

Eric D. Swain
U.S. Geological Survey, Miami, Florida, USA

Peter W. Swarzenski
U.S. Geological Survey, St. Petersburg, Florida, USA

Daniel W. Urish
University of Rhode Island, USA

Robb Whitaker
Water Replenishment of Southern California, USA

Contents

CHAPTER 1

Coastal Aquifer Planning Elements

M. Maimone, B. Harley, R. Fitzgerald,
H. Moe, R. Hossain, B. Heywood

1. INTRODUCTION

In many ways, groundwater resource planning in coastal areas requires an approach similar to more traditional water resource planning in inland areas. The same planning elements are common to both. Problems of aquifer yield, pumping interference, aquifer–stream interaction, and contamination from surface sources are all just as common along the coast as elsewhere, and just as difficult to solve. Aquifers situated along the coast, however, add a significant additional complication to the process of aquifer management: the potential for saltwater intrusion to eventually render portions of the coastal aquifer unusable as a source of drinking water.

This chapter focuses on the unique complication that potential saltwater intrusion poses for water resource managers in coastal areas, based on experiences gained over more than 20 years in the United States, Europe, and the Middle East. It discusses aquifer characterization, defines typical coastal aquifer problems, and outlines the basic steps for defining and evaluating potential management actions. It also discusses the use of saltwater intrusion models, without going into the detail. Additional material and some detail are available on the accompanying CD.

Figure 1 shows the recommended planning approach to coastal aquifer management. The sequence of planning elements, although made up of familiar elements, may appear to be in a somewhat unusual order. For example, the development of an integrated database is placed early in the sequence. This has proven to be an important step in making data analysis more effective, and in providing the necessary input for groundwater modeling in later phases. There are also two steps that focus on problem analysis and developing an understanding of the cause of elevated chloride

1-56670-605-X/04/$0.00+$1.50
© 2004 by CRC Press LLC

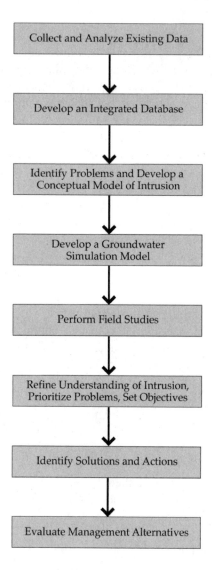

Figure 1: Study approach for coastal aquifer management.

concentrations. The first step is shown as the development of a theory for the cause of intrusion, often called a "conceptual model." It is during this step that stakeholder opinions and information is collected. The list of problems and issues should be revisited after modeling and analysis to finalize the list prior to setting planning objectives. In this latter step, problems perceived as important at the start of the study are reconciled with the results of modeling and analysis.

Probably the most unusual aspect of the recommended sequence shown in Figure 1 is the placement of the model development ahead of field studies. Preliminary modeling forces the planning team to develop an understanding of the data and a coherent theory of the mechanism of intrusion. This has been shown to significantly minimize the costs usually associated with extensive drilling and sampling by focusing the field study in areas most likely to yield important information. This approach is similar to the approach advocated by LeGrand [2000] that uses available piecemeal and imprecise information at the early stages of site studies through the use of conceptual modeling, generalizations, and inference to build a "prior conceptual model explanation" of the site phenomenon. LeGrand and the senior authors of this chapter have long advocated that hydrogeologic foreknowledge and preliminary modeling can often reveal more useful information than may be supplied by routine analysis. In our view, preliminary modeling should always precede the collection of extensive new data in the field.

Once field studies have been completed, it is expected that the preliminary model will be updated to reflect the additional data, and that the conceptual model of the mechanism of intrusion will be refined and confirmed by the field results. The final elements of the planning sequence include the identification of possible solutions to the intrusion problem (as well as other problems that have been identified) and the evaluation of the management alternatives.

In many cases, coastal aquifer planning is initiated by a water supplier, a regional government (e.g. a coastal county), or a state authority (national, or in the case of the USA, a state environmental agency). Planning usually starts because an intrusion problem has already occurred or is perceived to be a problem in the immediate future. Surprisingly, water suppliers and regulatory agencies have generally been slow to react to impending problems. Long range planning in coastal areas is still the exception rather than the rule. The cause of such inaction may be related to a lack of understanding of the mechanism of saltwater intrusion. In many coastal areas, such as along the Gulf of Mexico and the Atlantic coast of the USA, the onshore and offshore aquifer systems are highly stratified, with thick, confining units creating deep, confined aquifers. The existence of extensive, low permeable formations can result in large amounts of freshwater trapped in confined aquifers up to several miles offshore. For example, United States Geological Survey (USGS) studies have found freshwater beneath the ocean up to 50 miles off the Georgia and New Jersey coast [Kohout et al., 1988]. This represents a remnant of conditions from earlier ice ages, when the near-coast seabed was exposed during times of significantly lower sea levels. Although at the present sea level this water will naturally be replaced by saltwater, the process of migration of seawater back into the aquifer can take tens of thousands of years. Pumping along the coast, however, can accelerate the process significantly. What many coastal water suppliers fail to fully understand is that a significant portion of the freshwater they are withdrawing comes from this trapped, offshore freshwater. As water is withdrawn, it is replaced by saltwater. By pumping along the coast, they are, in essence, mining offshore freshwater. When this situation occurs, it is important for comprehensive coastal aquifer management programs to be put in place.

2. EXISTING DATA COLLECTION AND ANALYSIS

The first step in comprehensive coastal aquifer planning is to collect sufficient data to adequately define and understand the coastal aquifer system and its associated saltwater problems. Initially, existing data on aquifer heads and chloride concentrations in coastal wells should be reviewed. Usually data are sparse, with too few data points to adequately characterize or fully understand the current status of the aquifers with regard to saltwater intrusion. All data should be reviewed, including non-technical and anecdotal information, in addition to the more obvious physical or chemical data from supply wells or monitoring wells. An example of the possible importance of this data comes from a study on Long Island, where key information was provided by a homeowner whose well, long since abandoned, had gone salty in a particular year. This information provided an important piece of information that helped estimate the rate and direction of intrusion, and was one of the few "data points" for assessing the ability of the subsequently developed groundwater model to accurately simulate the historic pattern and rate of intrusion on the peninsula.

By examining and contouring heads along the coast, areas where offshore "mining" of freshwater often can be recognized. Heads in the freshwater aquifers may be below sea level, yet the wells continue to provide freshwater. Examples of this situation can be seen on Long Island in the deep, confined Lloyd Aquifer, and in Georgia and Florida, where suppliers take water from the confined Floridan Aquifer. Coastal water suppliers can often withdraw water from wells under these conditions for many years, even decades, before the offshore supply of freshwater is exhausted. However, once the tapped, offshore freshwater is depleted, the wells begin to withdraw saline water, and chloride concentrations usually rise rapidly to concentrations approaching those of seawater.

3. INTEGRATED DATABASE

Given the multi-disciplinary nature of coastal aquifer studies, one of the most important elements in the overall planning approach is adequate database development and application. Data must be organized in such a way that it can be analyzed spatially, in three dimensions, as well as temporally. As mentioned above, the long-term nature of interface movement requires that data from as far back as possible be collected. The only way to make the data available for analysis and modeling is to develop an integrated database/geographic information system (GIS). This critical, and often neglected, step of integrated database design allows users and modelers to analyze and query data, and places the data in a consistent format for model pre- and post-processing.

Data elements and map coverages in the database/GIS typically needed for coastal aquifer management include:

- Well information (depth, location, aquifer designation – even if preliminary)
- Historic and projected pumping information (linked to the well information)
- Chloride sampling data (dated, linked to well locations)
- Water level data (dated, linked to well locations)
- Surface map features (roads, streams, well locations, topographic features)
- Aquifer hydrogeologic parameters (transmissivity, hydraulic conductivity, formation thickness, specific yield, storativity, others). Data may exist as discrete points or spatial contours.
- Recharge estimates, mapped as contours if spatial variation is expected
- Maps of estimated present interface locations and depths

Long-term pumping records must also be collected. These data are critical to the development of a groundwater model. Unlike the calibration of a typical groundwater model in a freshwater aquifer, the response time of the freshwater/saltwater interface to changed pressure distribution (rise in sea level, increased pumping, altered recharge) in a coastal system might well be decades, or in some cases even a century or more. A critical part of the conceptual model is the estimate of the natural position of the interface prior to pumping, and a determination of whether the pre-development position was in equilibrium, or, as is common on the U.S. eastern seaboard, the aquifer is still responding to a long-term change in sea level from the last glacial period.

Due to the slow response of the interface, estimates of pumping rates over many decades must be made to test the model. Once the data have been put into a database/GIS, initial analysis can be carried out prior to modeling. Common analytical steps include examining:

- Water quality trend and spatial analyses
- Pumping analyses: monthly, seasonal, annual
- Water level and aquifer head mapping
- Chloride concentration and trend mapping
- Water demand projections

Most commercially available database software is now powerful enough to handle the data needs for even a large-scale regional aquifer management study. The key is to set up the database and the groundwater model in such a way that data can be moved from the database/GIS into the model, and model results can be transferred back to the database/GIS with relative ease.

4. IDENTIFY PROBLEM AND DEVELOP A CONCEPTUAL MODEL

Once available data and information have been collected and reviewed, a conceptual model of the mechanism of intrusion must be formed as a working hypothesis for further study. Intrusion generally can be categorized into one or more of several types of intrusion: horizontal and upward movement of the interface, downward leakage of brackish or saltwater from surface water (such as in estuarine environments), or saltwater upconing beneath a well field.

Horizontal intrusion, shown in Figure 2, occurs as the saline water from the coast slowly pushes the fresh inland groundwater landward and upward. This type of intrusion can be regional in scale, and results in the characteristic "wedge" of saltwater at the bottom of an aquifer. Its cause can be both natural (due to rising sea levels) and man-induced (pumping of freshwater from coastal wells). There is always an interface between the saltwater offshore and the freshwater onshore. This interface can sometimes be relatively sharp, with little or no transition or diffusion zone. Examples of this have been seen on Long Island, where vertical changes from seawater to freshwater have occurred over as little as 10 to 20 feet. In other cases, there may be a significant zone of transition. Note that there is always the potential for horizontal intrusion along the coast, and the interface is constantly shifting in response to sea level changes and changes in the freshwater aquifer head due to pumping or recharge changes.

Pumping from coastal wells can also draw saltwater downward from surface sources such as tidal creeks, canals, and embayments. This type of

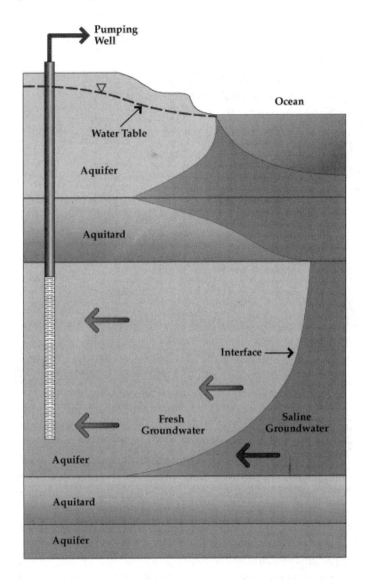

Figure 2: Horizontal saltwater intrusion toward a supply well.

intrusion, shown in Figure 3, is usually more local in nature. It typically occurs within the zone of capture of pumping wells where significant drawdown of the water table causes induced surface infiltration. This type of intrusion has occurred in areas of Florida, where drainage canals provide a means for saltwater to migrate inland. Another example in the USA is along the Delaware River, where saltwater moves up the river as river flows

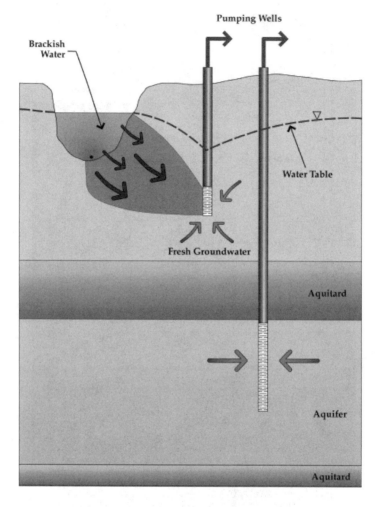

Figure 3: Induced downward movement of brackish surface water.

decrease, especially in drought conditions. In areas of New Jersey where groundwater pumping has induced recharge of river water from the Delaware River into the aquifer, saline water has contaminated portions of the aquifer near the river during periods of extended low flow in the river.

A third type of intrusion is called "upconing" and is shown in Figure 4. In this case, upconing occurs within the zone of capture of a pumping well, with saltwater drawn upward toward the well from saltwater existing in deeper aquifers or deeper portions of the same aquifer. This form of intrusion resembles an inverted funnel, hence the name "upconing." This is generally a more local intrusion problem, experienced by individual wells or well

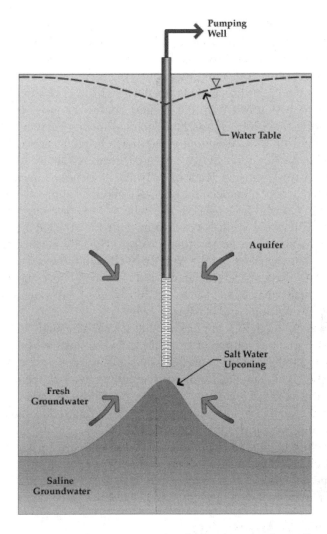

Figure 4: Saltwater upconing beneath a supply well.

fields. It requires that the pumping well be screened in freshwater overlying saltwater. At a certain pumping rate, a stable cone in the interface can develop below the well screen, but will not rise to the well. At increased rates of pumping, however, the cone can become unstable and the interface will rise abruptly toward and into the well, causing the well discharge to become saline. Upconing is a widespread problem, occurring in coastal areas such as in Florida, as well as in inland aquifers in Nebraska, Texas, New Mexico, and other states [Motz, 1992].

Although it might appear that the mechanism of intrusion would be obvious from the data, this is not always the case. An example of this is from a case study on Long Island. The Great Neck peninsula was one of the earliest peninsulas on the north shore of Long Island to be heavily developed as a suburb of New York City. By the 1930s, water supply pumping had caused the water table to decline by about 5 feet. As the demand for water grew from less than 1 million gallons per day (mgd) to over 4 mgd, more and more water was withdrawn from the deepest confined aquifer overlying the bedrock. During the 1960s, heads had declined in this deep aquifer (called either the Lloyd or Port Washington aquifer). In some areas, the United States Geological Survey (USGS) measured heads at 20 to 30 feet below mean sea level (msl), raising fears of saltwater intrusion. Elevated chloride concentrations were recorded in several public supply wells located approximately a mile from the shore, which were screened just above bedrock. The impacted wells were assumed to be affected by downward leakage of saltwater from nearby tidal creeks, primarily because an outpost well located between the public supply wells and the coast contained freshwater. The outpost well, screened only about 30 feet above bedrock, seemed to indicate that deep, horizontal intrusion of saltwater was not the cause of the closing of the public supply well.

However, modeling studies in the 1990s indicated that the cause actually was horizontal intrusion from the coast, followed by upcoming at the wells. The model results suggested that the saltwater wedge was so thin that the outpost well, screened only 30 feet above bedrock, continued to be screened in freshwater even as the saltwater moved below and past the well to the public supply wells. Subsequent drilling and downhole focused induction logging confirmed the modeling results.

Developing a well founded, conceptual model of the cause and mechanism of intrusion usually involves the interpretation of existing data, the development and use of a preliminary groundwater model, and the collection of additional data through field programs.

5. NUMERICAL MODELING

Although much insight can be gained from the process of collecting and analyzing the data, only through modeling of the mechanism of saltwater intrusion can the plausibility of the conceptual model be tested, and a deeper understanding of the mechanism of intrusion be gained. Modeling lies at the heart of the planning process, and interacts with all other activities as shown in Figure 5.

For this reason, it is recommended that a preliminary saltwater intrusion model be developed before additional field studies are carried out to collect more data. This is recommended for a number of reasons:

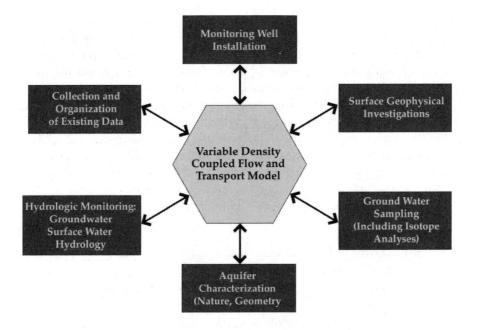

Figure 5: Investigation process.

- Models will provide significant insight into the potential mechanism of intrusion, and are the best tools for integrating and interpreting the data that is currently available.
- The model will provide clear guidance on the need for additional data, the type of data needed, and the most critical locations and depths to collect data.
- Models are the best tools for investigating and testing assumptions (e.g., assumptions of recharge, of interface location, etc.), developing and testing intrusion theories, and gaining an understanding of the sensitivity of the coastal aquifer system to changes in its hydrologic components (e.g., aquifer/aquitard structure, hydraulic conductivity, specific yield, etc.).
- Models are excellent visualization tools. Current modeling software packages now offer practical capabilities to zoom, pan, cut cross-sections through any part of the modeled area, contour heads or chloride concentrations in plan view or cross-section, show interface locations, and display point data in plan and cross-section.

For example, the study mentioned above in Great Neck started with an incorrect premise for the intrusion mechanism. In developing the preliminary model, it quickly became apparent that downward leakage of saline water was probably not the cause. It also became apparent that the current hydrogeologic data was insufficient, and that the published stratigraphy for the peninsula was probably not a good representation of the aquifer/aquitard system on the peninsula. The model became the tool to design the hydrogeologic investigation, locate and select well depths, and suggest areas for surface geophysical investigations. A well focused, cost effective field program was thus developed.

Selecting the correct model depends on setting clear and unambiguous modeling objectives. In many practical studies, modeling objectives could be to:

- Determine the cause of existing chloride contamination and the mechanism behind the contamination (lateral intrusion, upconing, downward leakage, etc.);
- Estimate the present, offshore location of the interface;
- Assess if the interface was stable prior to pumping;
- Determine the potential for intrusion or accelerated intrusion based on current pumping or future projected pumping;
- Estimate expected time of impact for specific well locations based on various pumping scenarios;
- Develop estimates of pumping rate versus rate of interface movement as part of a cost/benefit analysis of alternative water supply sources;
- Test various approaches to mitigating, halting, or reversing intrusion, or assess strategies for continued use of the aquifer as a viable water supply resource even with ongoing intrusion.

The modeling objectives, available budget, and the scale of the problem will be the primary factors in selecting an appropriate modeling approach.

One effective and practical approach to modeling the horizontal intrusion of saltwater is to apply some simplifying assumptions to enable reasonable but practical solutions that can quantify the relationship between salt- and freshwater, increase our understanding of the mechanism of intrusion, and make reasonable predictions about the response of the system to future conditions.

The most important assumption concerns the ability of the freshwater and the saltwater to mix. Under many coastal conditions, these two miscible fluids can be considered as immiscible, separated by a sharp interface or boundary. This assumption of a sharp interface has been used successfully in many studies, and significantly simplifies the mathematical formulation describing the physical process [Reilly et al., 1985].

Three-dimensional, sharp interface saltwater intrusion models are ideal tools to analyze the long-term sustainability of coastal wells in a regional context. These models can provide insight into the horizontal advance of wedges of saltwater under the influence of both sea level rise and coastal pumping. They can help estimate the rate at which freshwater is being withdrawn from offshore sources, and, provided that some information is available on the location of the offshore interface, can make accurate projections of the rate and timing of saltwater advance. In this way, the long-term viability of coastal well fields can be assessed, and plans for alternative sources or treatment of brackish water can be developed in a timely fashion. These models have formed the primary planning tool in areas of Florida and New York, and are often used to provide supporting evidence for applications for coastal water supply permits.

Another common means of simplifying computations is to model in two dimensions only. Two-dimensional cross section models may be useful for conceptual studies, e.g., estimating the historical rate of landward migration of an offshore seawater wedge due to the increase in sea level since the last ice age. Cross sectional models may also be useful for parameter estimation, e.g., estimating aquifer hydraulic parameters based on the head response in near shore wells to tidal sea level fluctuations. However, horizontal or "plan view" two-dimensional modeling is rarely appropriate for coastal modeling, and it should only be used with great caution. Even in simple, one-layer aquifer systems vertical flow is significant near the shoreline and must be explicitly considered. Failure to do so will yield incorrect results.

In analyzing upconing of saltwater, the existence of saltwater in aquifers below the pumping wells is usually already documented. In this situation, it is important to calculate the maximum sustainable pumping rate that still avoids saltwater upconing, or to calculate the timing of eventual upconing and the expected levels of chlorides in the wells. There are numerous analytical solutions to the saltwater upconing problem that can provide insight into the problem (see Motz [1992] and Schmorak and Mercado [1969]). Sharp interface models, fluid density-dependent flow, and coupled flow and transport models are also useful in simulating this situation in more complex hydrogeologic environments.

Single-phase contaminant transport models have proven very useful in analyzing the interaction between brackish surface water and groundwater where brackish surface water could be drawn downward toward pumping centers from canals, bays, or tidal creeks and rivers. In this case, the brackish water often has a density not significantly different from that of the groundwater. Advective transport and dispersion then become the primary mechanism of transport toward the well, a situation that can be effectively and efficiently simulated by single-phase transport and particle tracking codes.

If concentration gradients are important, then fluid density-dependent groundwater flow models, or coupled flow and transport models, can be applied. Fluid density-dependent models allow the effect on groundwater flow of fluid density gradients associated with solute concentration gradients to be incorporated into model simulations. The main applications are in studies of seawater intrusion where dispersion of salt into the freshwater zone needs to be quantified and mapped.

In the past 8 to 10 years, successful applications of fully three-dimensional models of saltwater intrusion, effective use of available analytical approximations of saltwater upconing, and the use of contaminant transport models have been combined to provide very effective planning and permitting tools for coastal water suppliers and regulatory agencies. These tools are particularly effective when fully integrated as a set of interrelated models.

Although the details of modeling are not the intent of this chapter, a number of observations about model calibration are appropriate. Unlike other types of groundwater models that can and should be carefully calibrated, saltwater intrusion models often do not have sufficient data to provide traditional "calibration and verification." Data gaps typically include a lack of data on the natural or equilibrium position of the interface, on whether the interface was in fact in equilibrium prior to pumping, on the current location and rate of movement, on the thickness of the saltwater wedge and the degree of diffusion in the transition zone, and on the history of intrusion (location, timing).

Because the intrusion is often a very slow phenomenon (intrusion rates of several feet per year are common), historical data with which to calibrate a model may be impossible to collect, even with an unlimited budget. This should not be a reason to forego modeling. Even with limited data, it is usually possible to test the "reasonableness" of the model results and learn a great deal about the coastal aquifer system.

One approach to establish parameter values and assess the accuracy of the model that has proven successful in many studies includes the following steps.

- Begin by developing a groundwater flow model of the freshwater system with offshore boundary conditions to represent the assumed present location of the interface. Calibrate as well as possible to the onshore head distribution in the standard fashion. Minimize the model error against measured heads. The resulting hydraulic parameters and spatial distribution from the freshwater model provide a reasonable starting point for the saltwater intrusion model. In some cases, little further adjustment of aquifer hydraulic parameters is required.

- Then apply the saltwater intrusion model (either a sharp interface or a coupled flow and transport model). Starting with the seawater offshore, conduct a long-term transient simulation to estimate the rate of seawater movement and equilibrium seawater position under pre-development (non-pumping) conditions.

- Use the model results to estimate the most likely predevelopment saltwater interface location. In a shallow aquifer system that is relatively unconfined, this could be onshore or just offshore. In such situations it is reasonable to assume equilibrium prior to the start of significant coastal pumping. In deeper, confined aquifers, the interface position may be onshore and in equilibrium. But it may also be offshore in equilibrium or, as along the U.S. east coast, still in transition toward equilibrium reacting to the approximately 300-foot sea level rise since the last ice age. In this case, it is particularly difficult to establish a starting position for the present day simulations.

- Determine the beginning time of significant pumping, and simulate the historical period of pumping with as accurate a representation of pumping stress as the data allow.

- Compare the simulated movement of the seawater with any existing historical data on chloride. In some cases, records exist of wells that went from fresh to salt at a certain date (rare but valuable data). In most cases, however, only a current, incomplete "snapshot" of the saltwater location can be assembled. Calibration can then only be done by assessing the reasonableness of the assumed predevelopment interface location, and simulating the movement of the interface over the historical record of pumping. The results should match the representation of the current position based on existing data. At the same time, adjustments to aquifer hydraulic parameters should be made as required so that the simulated head distribution agrees with measured water level data.

6. FIELD STUDIES

Having developed a preliminary model based on existing data, the gaps and inadequacy of the data is often apparent. At this point, field studies can be carried out to fill the most important data gaps. The design of a field study is very site specific. Several data collection techniques are briefly discussed in this section.

6.1   Well Drilling, Water Level Readings, and Chloride Sampling

The most direct approach (and often the most expensive) is to drill monitoring wells, preferably with the ability to measure chlorides at several depths. Drilling program objectives are commonly stated as:

- Providing sufficient coverage to accurately determine head distribution in the coastal area of interest,
- Collecting chloride concentrations to map the interface location,
- Gathering geological data to confirm or refute the initial conceptual model of the aquifer system and the mechanism of intrusion,
- Providing a permanent saltwater intrusion monitoring system. This should be enhanced by using PVC casing in the monitoring wells to allow downhole focused electromagnetic induction borehole geophysics to measure the thickness of the saltwater wedge.

In looking at chloride concentration results, it must be remembered that concentrations often change in the horizontal direction, with concentrations increasing toward the shore, but also in the vertical directions, with concentrations increasing with increasing depth. Care should be taken in mapping chloride distribution to account for variations in the depth of the sampling points.

One other consideration often overlooked is to correct mapped contours of aquifer heads for chloride concentration. Full seawater often has a specific weight of between 1.02 and 1.03, as opposed to freshwater, at 1.0. Thus a head of mean sea level with a chloride content of 19,000 ppm measured at a depth of 80 feet has an equivalent freshwater head of (+) 2 feet. In mapping the head contours near the coast, heads should be converted to equivalent freshwater heads. If this is not done, flow direction can be completely misinterpreted. The equation below [Lusczynski and Swarzenski, 1966] can be used to convert heads of saline or brackish water to equivalent freshwater heads:

$$p_f \, H_{if} = p_i \, H_i - Z_i \left(p_i - p_f \right) \tag{1}$$

where p_f is the density of freshwater
  H_{if} is the equivalent freshwater head of water in well (brackish or saline) at point i
  p_i is the density of the saline or brackish well water
  H_i is the head measured in the well (brackish or saline)
  Z_i is the elevation of point i (measured positively upward from the screen elevation or the depth of measurement).

6.2   Chloride Balances and Ion/Isotope Fingerprints

A chloride balance (estimating the mass of chloride from each potential source, and comparing it to the mass measured in the aquifer of concern) is another useful field study that can help in the investigation of potential sources of contamination. In many coastal aquifers, the only significant source of chloride contamination of the aquifer is seawater intrusion. In such cases, chloride balances serve no real purpose. In certain cases, however, the collection of data and the modeling might reveal several potential sources. For example, in recent studies of the Gaza coastal aquifer, chlorides were noted to be impacting the shallow aquifer system along the coast. One obvious potential source was lateral intrusion of seawater; however, there were other potential sources as well:

- Brackish water in the same aquifer further inland as a result of naturally occurring minerals in the aquifer material;
- Recharge of concentrated wastewater containing high TDS (total dissolved solid) and salt concentrations from septic systems, sewage infiltration, and agricultural irrigation water;
- Upconing of deep brines from the underlying aquifer.

When faced with multiple potential sources, developing a chloride balance can yield significant insight into the relative importance of each source. Sampling and developing diagrams of the relative concentrations of ion in the water of each potential source can help to "fingerprint" each source. Comparing the fingerprint of the contaminated groundwater with the fingerprint of each source provides another way to identify contamination sources.

6.3   Surface Geophysical Studies

One proven survey approach is the use of Time Domain Electromagnetic (TDEM) soundings. This technique is effective because electrical resistivity is highly influenced by the salinity of the groundwater, providing clear contrast between zones of freshwater and zones of saltwater. In general, TDEM has proven to have excellent vertical and lateral resolution for mapping interfaces characterized by resistivity contrasts, and can reach depths of between 500 and 1000 feet, depending on the availability of open space at the surface.

In electrical and electromagnetic techniques for measuring electrical resistivity in the ground, resistivity is measured by determining the resistance to flow of electrical current. TDEM currents are induced by a time varying magnetic field of a transmitter, in this case a loop of insulated wire laid on the ground. Transmitter loop sizes of 100 by 100 feet and 200 by 200 feet are common, depending on the depth of exploration required. A multi-turn air coil receiver is placed in the center of the loop. The receiver measures the electromotive forces due to the secondary magnetic field caused by the subsurface currents. The measurements reflect the subsurface resistivity at increasing depth with increasing time. The objective of processing the TDEM data is to obtain a "solution" for the resistivity profiles obtained that best fits the data and reflects both the stratigraphy and the salinity profile of the subsurface. Resistivity mainly depends on porosity, salinity of the pore water, and the clay content of the formation.

TDEM surveys are relatively inexpensive, and can provide many data points with which to map both the location and depth of the interface. With limited, follow-up well drilling, the results can be confirmed, and correlations between salinity and the resistivity readings can be developed.

6.4   Downhole Focused Induction Logging

Focused-induction logging of boreholes uses an electromagnetic emitter coil that induces current loops within the surrounding formation to generate a secondary electromagnetic field. The intensity of the secondary field received by the receiver coil is proportional to the formation conductivity [Stumm, 1993]. Saltwater, in place of freshwater, significantly alters the conductivity, and is usually easily recognized in the downhole log of the well. The log provides an excellent indicator of the exact depth of the transition from freshwater to saltwater in the well. The advantage of focused induction logging is that it measures conductivity at a fixed distance from the borehole, and thus can be set up to measure pore water outside the well, if the well is uncased or is cased with PVC.

This technique can be used to locate the exact depth of the interface by drilling a well through the overlying freshwater into the underlying saltwater. The logs can then show the exact depth and thickness of the transition zone. The wells can also be used as permanent monitoring stations to measure the rise of the interface over time in areas where the interface is moving. Annual logging over a period of years can provide an important record of interface movement.

7.   REFINE AND PRIORITIZE PROBLEMS AND SET OBJECTIVES

Once the analyses and modeling have been carried out, there should be a more definitive understanding of the problem and a refined conceptual model. The conceptual model and understanding of the problem will have been either verified or further refined by the field studies. To move toward a set of solutions, however, the problems need to be turned into a clear set of planning objectives.

The model can provide a practical tool with which to understand the severity and time scale of the problem and to present a clear set of potential planning objectives to a stakeholder group. Examples of possible model results might be:

- The current pumping does not cause significant movement of the interface.
- The interface is already onshore, and relatively stable with regard to further onshore movement.
- The intrusion is an upconing problem beneath specific wells. The cone might be unstable, and chlorides could enter the well above a certain critical pumping rate in a matter of days or weeks. Alternatively, the interface depth might be such that while upconing might eventually take place, the process could take 5 to 10 years.
- The interface is at an unknown, offshore location. Pumping is significant and will move the interface onshore and impact wells; however, eventual impacts through horizontal interface movement might take decades.
- The interface is onshore and adequately mapped, and is continuing to move toward pumping wells, and impacts can be expected in a number of years.

At this point, it is important to actively engage the technical advisory committees or other stakeholder groups in the planning process. The goal is to gain consensus on the nature and severity of the intrusion problem, and to develop a set of operational planning objectives. Some examples may help to clarify this point.

- It might be found that mining of offshore freshwater is occurring. A decision must be made whether this is acceptable for the present, or whether the goal is to halt or even reverse intrusion. Either solution might be appropriate, depending on the availability of alternative sources, the cause of the intrusion, the feasibility of solutions, and the priorities of the stakeholders and decision makers. If impacts are estimated to occur 50 to 75 years in the future, one objective might be to accept offshore mining of water for 25 years, and then to slowly shift toward alternative sources by year 50. Alternatively, a regulatory agency might decide to impose strict pumping limitations on all coastal well fields.
- It might be determined that the interface is presently very close to pumping wells, and that present coastal pumping is close to the sustainable yield. One objective might be to reduce pumping and halt all further intrusion. Alternatively, if water is scarce, this might not be

a viable objective, and the objective might be to maximize the aquifer yield by finding the most productive rate and distribution of wells to continue to extract freshwater without causing direct impacts to wells.

The final outcome of this stage of planning should be a clear set of objectives for which a set of solutions or actions can be developed to meet the planning objectives.

8. IDENTIFY SOLUTIONS

Once the planning objectives have been identified, potential means to mitigating intrusion can be investigated. Examples of potential solutions include:

- Enhanced Aquifer Recharge: increase freshwater heads to resist seawater intrusion by spreading surface water, or treated wastewater, or by capturing surface runoff in recharge basins.
- Demand Management: essentially lowering the demand for water to reduce pumping stresses on the aquifer.
- Non-Potable Water Reuse: another method to reduce demand is by replacing potable water with treated wastewater for irrigation or other non-potable water uses.
- Injection Barriers: a hydraulic barrier can be created by injecting water to form a narrow zone in the aquifer in which the freshwater gradient is toward the sea. This prevents intrusion of seawater into unaffected portions of the aquifer system.
- Extraction Barriers: a seldom used solution that creates an intrusion barrier by extracting saline water near the shoreline, thus protecting wells further inland.
- Tapping Alternative Aquifers: aquifers located either below or above the impacted aquifer can sometimes provide alternative sources and relieve pumping stress on the impacted aquifer.
- Well Relocation: relocating wells to areas of higher freshwater head or areas less susceptible to intrusion. Relocation can also be used to reduce the intensity of pumping in an area and spread out the pumping cone of depression, making the head gradients less steep, and reducing the potential for localized intrusion.
- Plugging Abandoned Wells: in some cases, older abandoned wells are left in place and can provide a conduit for leaking saltwater from saline aquifers into fresh aquifers.
- Modified Pumping Rates: in situations where a production well is subject to periodic increases in salinity due to upconing, a modified

pumping schedule (lower constant rate or an on–off sequence that allows the well heads to recover) can sometimes alleviate the problem.

- Pumping Rate Caps: restrictions on pumping rates or on the placement of new wells can be applied in intrusion-sensitive areas to protect against additional intrusion.
- Physical Barrier: physical barriers such as slurry walls or sheet piles have been tried in small-scale, shallow intrusion situations to protect a well. This is not a common approach.
- Scavenger Wells: these are very shallow wells specifically designed for extracting freshwater while preventing the upconing of saline water through hydrodynamic stabilization of the saline–freshwater interface. These systems pump freshwater from the upper part of the aquifer; in some cases simultaneously pumping and wasting saline water to produce a zone where freshwater will collect for extraction.
- Controlled Intrusion: mining trapped offshore freshwater for use, with adequate planning for alternative supplies when the source is depleted.
- Intrusion with Treatment: saltwater intrusion could be tolerated at certain concentrations, with treatment to remove the salinity before use for public supply. This could range from limited treatment (by reverse osmosis) of brackish water, to full desalination plants using groundwater with close to seawater salinity.
- Conjunctive Use: the coordinated use of surface water supplies and storage with groundwater supplies and storage to offset excessive reliance on groundwater.
- Aquifer Storage and Recovery: the temporary storage of potable water in a saline aquifer for later extraction and use. These systems may be useful in managing peak seasonal demands, but they will not provide long-term management of unsustainable average demands on a coastal aquifer.

9. EVALUATE MANAGEMENT ALTERNATIVES

An important aspect of coastal aquifer planning is the selection of alternative solutions, all of which typically involve making complex tradeoffs. Ideally, the selection should be based on a thorough evaluation of competing alternatives in an organized, comprehensive, and defensible manner. Multi-criteria evaluation techniques have proven to be an excellent decision support tool for evaluating water resource management alternatives.

Much research has been done on multi-criteria evaluation techniques with the aim of developing simple, understandable, yet effective decision support tools. Approaches include simple weighted summation matrix techniques, concordance–discordance analysis, GIS overlay techniques, and mixed data multi-criteria techniques. Each of these procedures attempts to include economic, environmental, social, technical, political, and other considerations within the decision making process. Given the complexities involved in the effective management of coastal aquifers, use of such tools are essential to the development of practical and implementable aquifer management solutions which will meet multi-faceted community needs.

An example of the challenges faced in a coastal aquifer system is that presently being experienced in the Gaza coastal aquifer in Palestine. The population in Gaza has been increasing rapidly and the area is almost totally dependent on the underlying coastal aquifer for its water needs. Presently, there are severe shortages of drinking water, and very serious concerns about the quality of the available water. Over-exploitation of the aquifer during the past 20 years has led to rapid and growing seawater intrusion into this sole-source aquifer system. The Palestinian Water Authority (PWA) has to manage two major competing uses for the water: drinking water for the highly populated cities and towns, and irrigation water for traditional agricultural activities. Their goal in a recent aquifer management study was to develop a comprehensive plan for water resources management that identified a "preferred" set of aquifer development schemes that would then be presented for funding by international agencies. The complexity of the decision process for the Coastal Aquifer Management Plan (CAMP) lay in the sheer number of options possible, and the difficulty in combining them into rational, defensible, and optimal combinations to form a comprehensive long-range plan.

The approach used in this instance was to create groups of options that each had a similar objective (comparing "like" things). In the Gaza plan, there were two sets of option groups: those designed to increase the overall availability of water (quantity), and those schemes that would primarily improve water quality. In all, over 100 options were identified in the early planning phases for further evaluation. The groups of water "quantity" options, and a few representative "schemes" within each group are presented in Table 1 to illustrate the complexity of the problem. A sophisticated, but spreadsheet based, multi-criteria evaluation program, EVAMIX [Voogd, 1982], was used in the development of the CAMP to organize a complex decision process, and to support a high level planning group in making key water resource decisions for this highly stressed coastal aquifer.

The evaluation of each of the groups of options used evaluation criteria selected from the following seven categories of impacts, depending on the relevance to the option group:

QN-1: Alternative Source Options

- Seawater desalination using membrane technology
- Import water by sea in tankers from Turkey
- Import water from West Bank via a pipeline

QN-2: Agricultural Demand Options

- Reduce agriculture production, increase food import
- Increase irrigation efficiency through drip irrigation
- Improve water quality to reduce soil flushing requirements

QN-3: Domestic Demand Options

- Increase pricing to suppress demand
- Apply vigorous water conservation measures
- Separate drinking/cooking water from other domestic use

QN-4: Commercial/Industrial Demand Options

- Stimulate the development of low water use industry
- Aggressively apply water conservation measures

QN-5: Water/Wastewater Distribution Options

- Build separate freshwater/brackish water lines to homes
- Timed freshwater delivery for a few hours each day
- Reduce system loss by leakage
- Build separate treated wastewater distribution system (including storage capacity) for agriculture
- Develop schemes to recharge wastewater

QN-6: Wastewater Collection Options

- Collection and pumping to centralized treatment plants
- Separation of stormwater and wastewater to permit recharge of stormwater

Table 1: Water quantity option categories and example management schemes, Gaza Coastal Aquifer management study.

QN-7: Wastewater Treatment Options

- Secondary treatment plus recharge for indirect reuse
- Tertiary treatment (membrane desalination) of saline groundwater to meet unrestricted agricultural use standards

QN-8: Wastewater Reuse Options

- Develop direct reuse by agricultural users
- Conjunctive use: agricultural in dry season, recharge in wet season
- Build a North–South pipeline for regional distribution of treated wastewater
- Trade treated wastewater to Israel for freshwater

QN-9: Pumping Management Options
- Full Palestinian Water Authority control of all aquifer pumping
- More limited Palestinian Water Authority control of public supply and regulated agricultural pumping

QN-10: Enhanced Stormwater Collection Options
- Centralized collection and recharge using large basins
- Dispersed collection and recharge through drywells and perforated pipes
- Household collection and storage in cisterns

Table 1: (Continued)

1. Financial and Economic Impacts
2. Technical Considerations
3. Source Viability
4. Political Considerations (both local and international)
5. Institutional Considerations
6. Environmental Impacts
7. Social Impacts

Weighting factors were assigned to each of the evaluation criteria to represent their relative importance to the individual stakeholders and decision makers. It is essential that these weighting factors, either objective or subjective, be allowed to be specific to each stakeholder (or group). Only by doing this will all the groups be fully vested in the resulting plan.

In the Gaza situation, the planning process involved evaluating each of the 17 (10 Water Quantity and 7 Water Quality) option groups separately. This was done in a facilitated 2-day workshop involving the primary decision makers from the Palestinian Water Authority (PWA), the Ministry of Planning, the Ministry of Environment, and other interested parties. Prior to the meeting, the expected impacts of each of the various options was simulated and evaluated using a three-dimensional variable-density flow and solute transport groundwater model of the multi-layer regional aquifer system. These estimated impacts, which addressed issues such as long-term recovery in aquifer piezometric heads, the growth or shrinkage of the seawater intrusion lenses under various pumping schemes, the impact of proposed additional municipal wells, the response to stormwater recharge, etc., were key inputs that were available to the workshop participants.

As the participants evaluated the plan elements, decision matrices were projected on a screen, and initial criteria weights were applied in an interactive process. The EVAMIX process then computed the relative rankings of the options, and stimulated discussion of the results among the participants. Criteria weights were often varied by individual participants to test the sensitivity of the evaluation process. Such re-assessment usually stimulated further discussion. In many cases, the results helped to focus the debate to such a degree that consensus was reached on certain plan options without further analysis. Over the course of the workshop, the most promising and globally "acceptable" technologies within each of the 17 groups were identified. These options were later assembled into an overall plan of action (the "CAMP"), which was adopted by the PWA as their long-range water resources master plan.

Use of a multi-objective evaluation tool during the development of the CAMP—for decision support as well as for improved stakeholder involvement—resulted in a well defined, transparent, and defensible planning approach. Although debate at the planning workshop was often lively, the group invariably reached a consensus on the ranking of the alternatives within a reasonable amount of time. The resulting plan (the CAMP) incorporated and integrated modeling results, stakeholder view points, policy factors, and socio-economic considerations into a comprehensive long-term plan that will allow continued development of the Gaza Strip, while at the same time both protecting and maximizing the effective use of the underlying coastal aquifer.

10. SUMMARY

Coastal aquifers present very complex and unique management challenges. Their effective management requires balancing many competing demands, and typically requires the use of a suite of numerical models, field investigations, and the development of a consensus on proposed management options by many levels in state, local (and sometimes federal or central) governments, and other concerned groups. The time scales for experiencing the impacts of management decisions may be relatively long, but if these critical coastal resources are adversely impacted by over-development and consequent seawater intrusion, remedial measures are at best very difficult and expensive to implement. In many instances seawater-impacted coastal aquifers cannot be restored to a viable freshwater condition.

REFERENCES

Blackhawk Geosciences, "TDEM Survey for Delineation of Salt Water Intrusion, Great Neck Peninsula, Long Island, New York", Engineering Report to NCDPW, 1990.

Camp Dresser & Mckee, "Potential Salt Water Intrusion at Public Water Supply Wells in Great Neck, New York", Engineering Report to NCDEPW, 1992.

Kilburn, C., "Hydrogeology of the Town of North Hempstead, Nassau County, Long Island, N.Y.", *USGS Water Resources Bulletin*, **12**, 1979.

Kohout, F.A., Meisler, H., Meyer, F.W., Johnston, R.H., Leve, G.W., and Wait, R.L., "Chapter 23: Hydrogeology of the Atlantic continental margin", The Geology of North America, Vol. 1–2, The Atlantic Continental Margin: US, The Geological Society of America, 1988.

LeGrand, H.E. and Rosen, L., "Systematic Makings of Early Stage Hydrogeologic Conceptual Models", *Ground Water*, **38**(6), 887–893, 2000.

Lusczynski, N.J. and Swarzenski, W.V., "Salt-Water Encroachment in Southern Nassau and Southwestern Queens Counties, Long Island, New York", USGS Water Supply Paper 1613-F, 1966.

Maimone, M., Keil, D., and Hoekstra, P., "Geophysical Surveys for Mapping Boundaries of Fresh Water and Salty Water in Southern Nassau County, Long Island, New York", *3rd National Outdoor Action Conference Proceedings*, NWWA, Orlando, Florida, 1989.

Motz, L.H., "Salt Water Upconing in an Aquifer Overlain by a Leaky Confining Bed", *Ground Water*, **30**(2), 1992.

Reilly, T.E. and Goodman, A.S., "Quantitative Analysis of Saltwater-Freshwater Relationships in Groundwater Systems—a Historical Perspective", *Journal of Hydrology*, **80**, 125–160, 1985.

Schmorak, S. and Mercado, A., "Upconing of Fresh Water–Sea Water Interface Below Pumping Wells, Field Study", *Water Resources Research*, **5**(6), 1969.

Stumm, F., "Use of focused electromagnetic induction borehole geophysics to delineate the saltwater-freshwater interface in Great Neck, Long Island, New York", in Proceedings, Symposium on the Application of Geophysics to Engineering and Environmental Problems, v. 2. p. 5132–525, 1993.

Voogd, H., "Multi-Criteria Evaluation with Mixed Qualitative and Quantitative Data", *Environment and Planning*, **9**, 221–236, 1982.

CHAPTER 2

Saltwater Intrusion in the Coastal Aquifers of Los Angeles County, California

T.A. Johnson, R. Whitaker

1. INTRODUCTION

The Central and West Coast groundwater basins (CWCB) are two coastal aquifer systems located adjacent to the Pacific Ocean in southwestern Los Angeles County, California (Figure 1). Severe groundwater overdraft of these basins from the early 1900s to the late 1950s caused water levels to drop below sea level, allowing saltwater to intrude into the potable aquifers, knocking coastal wells out of service, and threatening the usability of this major water supply reservoir.

In an effort to halt the intrusion and control the overdraft, groundwater management agencies took three major steps from the mid-1950s to mid-1960s, including 1) construction of freshwater injection wells along the coast to prevent the saltwater intrusion by the Los Angeles County Flood Control District (LACFCD); 2) seeking adjudication of the groundwater basins to limit the amount of groundwater that could be pumped annually; and 3) creation of the Water Replenishment District of Southern California (WRD) to purchase artificial replenishment water to make up the annual and accumulated overdrafts, purchase barrier injection water, and to protect the water quality of the CWCB.

This chapter will describe the current saltwater barrier system in the CWCB, the monitoring, modeling, and management efforts underway to prevent continued intrusion, and present the results of an investigation to identify cost-effective alternatives to injection wells for saltwater intrusion control. Copies of several reports that provide additional information about the groundwater basins and the barrier alternative study are available on the accompanying CD.

1-56670-605-X/04/$0.00+$1.50
© 2004 by CRC Press LLC

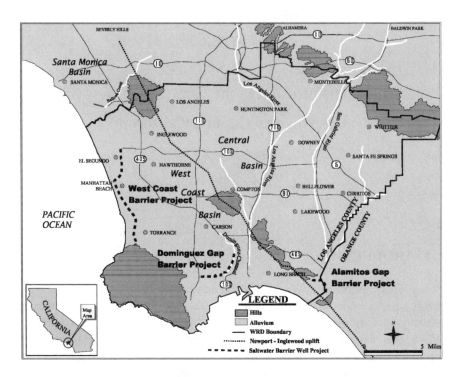

Figure 1: Study area and location of saltwater barrier projects.

1.1   Project Area

The coastal plain of Los Angeles County contains four groundwater basins, including the Central Basin, the Hollywood Basin, the Santa Monica Basin, and the West Coast Basin. The saltwater intrusion barrier well projects exist in the Central Basin and West Coast Basin.

The Central Basin is bounded on the north by the Elysian, Repetto, Merced, and Puente Hills, on the east by the Los Angeles County/Orange County political line, and on the south and west by the Newport-Inglewood uplift, which is a series of en-echelon faults and folds that act as partial to full barriers to groundwater flow.

The West Coast Basin is bounded on the east by this uplift structure, on the south by the Pacific Ocean (San Pedro Bay) and the Palos Verdes Hills, on the west by the Pacific Ocean (Santa Monica Bay), and on the north by the Ballona Escarpment, which is not a structural feature but the approximate location of a groundwater divide [California Department of Water Resources (CDWR), 1961].

The CWCB overlie an area of about 420 square miles and 4 million people, and include 43 cities such as Torrance on the east, Cerritos on the west, Whittier and a portion of Los Angeles to the north, and Long Beach to the south. Total water demand by the people and businesses in this area is about 730,000 acre-feet per year (afy). About 255,000 afy of this demand is met by local groundwater production (35%) from over 400 production wells. Approximately 440,000 afy is imported into the region from northern California and the Colorado River (60%), and the remaining 5% is locally treated and reused wastewater [Water Replenishment District of Southern California, 2001].

1.2   Hydrogeology

The CWCB are comprised of Quaternary alluvial and marine sedimentary deposits layered into permeable multiple aquifer systems comprised of fine to coarse sand and gravel, and less permeable aquitards comprised of fine sand, silt, and clay. They are predominantly confined aquifers, but are semi confined to unconfined in the northern forebay areas. Thickness of the basin varies across the coastal plain, but typically range from several hundred feet thick to more than 2,000 feet thick due to structural faults and folds that cross and shape the basins [CDWR, 1961].

The practical base of the groundwater basin is the contact with the underlying Pliocene Pico Formation, which is comprised of marine silts and clays with occasional interbeds of sand and gravel. Although the Pico Formation can sometimes provide minor amounts of water to wells, the vast majority of groundwater is produced from the overlying Quaternary sediments and therefore constitutes the main aquifer system. Details on the geologic history, physiography, stratigraphy, geologic structure, and groundwater basins can be found in CDWR [1961] and Reichard *et al.* [2002].

The principal geologic formations that contain the aquifer systems within the project area include, from shallowest to deepest, the Recent alluvium (Gaspur Aquifer), the Upper Pleistocene Lakewood Formation (Exposition, Artesia, Gardena, and Gage aquifers), and the Lower Pleistocene San Pedro Formation (Hollydale, Jefferson, Lynwood, Silverado, and Sunnyside aquifers). In the West Coast Basin, the Gage Aquifer is also known as the "200-foot sand" Aquifer, the Lynwood is known as the "400-foot gravel" aquifer, and the Sunnyside is referred to as the Lower San Pedro Aquifer. Figure 2 shows the relationship between geologic formations and aquifers in the CWCB. Along the coast, these aquifers can extend offshore allowing potential pathways for saltwater intrusion to occur.

Movement of the groundwater in the CWCB is from areas of recharge to the areas of discharge. The main areas of recharge are from spreading basins located in the northeast portion of the Central Basin, which

AGE	FORMATION	AQUIFER*	AQUIFER SYSTEM
Holocene	Active Dune Sand	Semi-Perched	Recent Aquifer System
		Gaspur	
Upper Pleistocene	Older Dune Sand	Ballona	Lakewood Aquifer System
	Lakewood	Exposition	
		Artesia	
	Formation	Gage / Gardena (200 Foot Sand)	
Lower Pleistocene	San Pedro Formation	Hollydale	Upper San Pedro Aquifer System
		Jefferson	
		Lynwood (400 Foot Gravel) [C,B,A,I Zones]	
		Silverado [Main]	
		Sunnyside/ Lower San Pedro [Lower Zone]	Lower San Pedro Aquifer System
Upper Pliocene	Pico Formation		

* Names in parenthesies () are West Coast Basin synonyms for Central Basin aquifers.
  Bracketed names [ ] are Orange County equivalent aquifer names.
  Modified from CDWR (1961), Reichard, et. al (2002), and Lipshie and Larson (1995)

Figure 2: Geologic formations and aquifers, central and west coast basin.

infiltrate locally derived and captured stormwater as well as artificial replenishment water which is purchased to make up the overdraft. Significant recharge also occurs through injection of imported and recycled water into the saltwater barrier wells. Other recharge components include groundwater underflow from adjacent basins, infiltration of precipitation and surface applied water, and continued saltwater intrusion in some areas. The most significant discharge from the CWCB is groundwater extractions that nearly equal all of the inflow components. Groundwater underflow is also an outflow component. During a modeling base period from 1971 to 1996, it was reported by the Water Replenishment District of Southern California [2001] that inflows averaged 252,500 afy (natural inflows 141,600 afy and

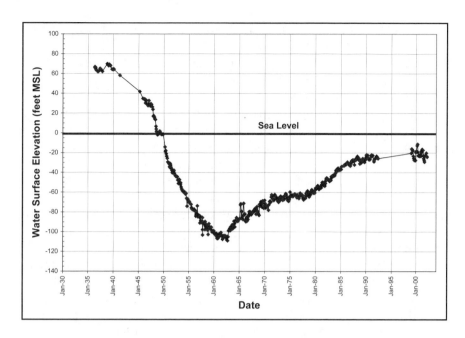

Figure 3: Hydrograph of water well 2S/13W-10A1.

artificial inflows 110,900 afy), and outflows averaged 253,300 afy (250,700 afy groundwater pumping, 2,600 afy groundwater underflow).

1.3   Groundwater Overdraft and Saltwater Intrusion

As populations grew and industrial and agricultural development advanced in the early half of the 20th century, groundwater extractions outpaced natural replenishment causing severe overdraft. Although the safe yield of the CWCB was estimated at 173,000 afy by the CDWR [1962], groundwater production was more than double. This resulted in a rapid lowering of groundwater levels to below sea level in some cases (Figure 3), creating the opportunity for saltwater to intrude inland because of the landward-induced gradient.

The probability of saltwater intrusion into the coastal aquifers of Los Angeles County was first predicted by the United States Geological Survey (USGS) in the early 1900s [Mendenhall, 1905]. Mendenhall mapped the artesian areas in coastal Los Angeles County, and noted that their area was declining. Callison and Roth [1967] describe that degradation of groundwater quality due to saltwater intrusion was first reported in Redondo Beach in 1912, in Hermosa Beach in 1915, and in El Segundo in 1921. The California Department of Water Resources [1962] estimated that up to 600,000 acre-feet of saltwater intruded and contaminated the Los Angeles

Coastal aquifers by the late 1950s. Reports by Poland *et al.* [1959] and the California Department of Water Resources [1950] helped define and quantify the problem, which was recognized as a serious threat to the water resources of the growing Los Angeles area.

To address the declining groundwater levels and loss of groundwater from storage, and to minimize the saltwater intrusion, groundwater management agencies implemented three important measures from the mid 1950s to the mid 1960s. It was recognized that serious groundwater overdraft was occurring as pumping exceeded natural recharge. Therefore, lawsuits were filed and the courts set a limit (adjudication) on the amount of groundwater that could be pumped from the CWCB. The West Coast Basin adjudication took effect in 1961 and capped production to rights holders in the amount of 64,468 afy. The Central Basin adjudication took effect in 1965, limiting pumping to an allowed pumping allocation of 217,367 afy. This total amount of 281,835 afy, however, still exceeded the natural replenishment of the basins.

In 1959, the WRD was created through a special election in Los Angles County to manage artificial replenishment in the CWCB and to make up any overdraft. WRD determines the annual overdraft annually, and purchases imported and recycled water for replenishment at spreading grounds and through the saltwater barrier wells. Excess replenishment water has also been purchased to help make up the accumulated overdraft. In water year 2000/2001, WRD purchased 113,913 af of artificial replenishment water [WRD, 2002]. Since 1959, over 255,000 af of groundwater has been returned to storage [WRD, 2002]. Details of the groundwater conditions and artificial replenishment activities in the CWCB can be found in WRD's Regional Groundwater Monitoring Report for Water Year 2000/2001 and the Engineering Survey and Report 2001 that are contained in the accompanying CD of the book.

The third significant event to mitigate the CWCB overdraft problems was the implementation and construction of the saltwater barrier projects, which are described in detail below in Section 2. The net effect of these three management implementations was a stoppage to the overdraft, an increase in water levels and groundwater in storage, and a halt to saltwater intrusion. This reversal effect can be seen in the hydrograph in Figure 3, where in the early 1960s there was a dramatic reversal in water levels in this well (and wells throughout the CWCB) from a declining trend to a rising trend.

2. SALTWATER BARRIER PROJECTS

In the early 1950s, the LACFCD undertook testing to evaluate the use of injection wells for saltwater intrusion control. In 1951, using an abandoned water well in Manhattan Beach, an injection test was conducted where a freshwater mound was established and successfully maintained in a confined aquifer [Lipshie and Larson, 1995]. That test lead to a larger test project located in the cities of Manhattan and Hermosa beaches. A line of 9 recharge wells, spaced 500 feet apart, and 54 observation wells were constructed and used for the test. Treated Colorado River water was used for the injection source, and injection and monitoring occurred between February 1953 and June 1954. The test successfully created a pressure ridge along the injection line, reversing the previous landward gradient that allowed the intrusion. The results of the test project are well documented in the CDWR [1957].

Based on the success of the project, a cost benefit analysis, and after evaluating other alternatives including a puddled clay-filled trench, basin wide reduction of pumping, direct recharge through spreading basins, creation of a pumping trough parallel to the coast, and emplacement of a grout cutoff wall [Callison and Roth, 1967], the LACFCD expanded the injection well system over the next 20 years into three separate barrier projects stretching over 16 miles of coastline. The three barrier projects are the West Coast Basin barrier project, the Dominguez Gap barrier project, and the Alamitos Gap barrier project (Figure 1). These projects require an extensive infrastructure of injection wells, observation wells, extraction wells, over 50 miles of pipeline to carry freshwater to the injection wells, pipelines for the disposal of saline water from extraction wells, water pressure reduction stations, and various electrical distribution centers that provide the power for the pumps. Nearly the entire infrastructure is underground, minimizing the impact to the overlying and heavily urbanized Los Angeles region.

The wells inject freshwater into the principal aquifer systems, including the Gage, Lynwood, Silverado, and Sunnyside. Because of the multiple aquifer systems, the LACFCD often constructed the wells with the ability to inject into different aquifers at different rates from the same well. This was accomplished using a packer system to isolate upper from lower zones.

The wells inject over 30,000 acre-feet per year of both potable water and highly treated recycled water into the CWCB aquifers year round. Regulatory agencies require that the injected water be very high quality so as not to degrade the quality of the drinking water aquifers. Table 1 provides a

Barrier Project	West Coast	Dominguez Gap	Alamitos
Date Begun	1953	1969	1964
Length (miles)	9	4.3	2.2
Injection Wells (by 2003)	153	94	44
Extraction Wells	0	0	4
Observation Wells	276	232	239
Distance from Coast (miles)	1	0.5-2.8	2
2000/2001 Injection (af)	20,826	3,923	5,633
2000/2001 Injection Water Costs	$10,300,000	$2,000,000	$2,500,000
Approximate Maintenance Costs	$2,500,000	$1,000,000	$1,000,000

Table 1: Summary of barrier project information.

summary of the barrier projects. Typical well construction diagrams for a dual recharge well and a single recharge well are shown as Figure 4. Details of the three barrier systems are presented in the following sections.

2.1 West Coast Basin Barrier Project

The West Coast Basin Barrier Project (WCBBP) was the first of the three barrier projects. It was designed and constructed to protect the western coastline of the West Coast basin from saltwater intrusion. It was begun in 1951 as part of the initial pilot testing by the LACFCD and completed in 1969. Several additional monitoring wells were drilled in 1995. The WCBBP consists of 153 injection wells and 276 observation wells extending along a 9-mile stretch from the Palos Verdes Hills northward to the Los Angeles International Airport. The barrier alignment is about 1-mile inland and parallel to the coastline.

Pathways for intrusion in this area are through the Gage Aquifer (200-foot sand), Silverado Aquifer (which is merged with the Lynwood/400-foot gravel in this area), and the Sunnyside Aquifer (Lower San Pedro) that extend offshore and are direct conduits for the saltwater. The injection wells are completed into these three aquifers at depths from 250 feet to 700 feet below the ground surface, with an average depth of about 450 feet. The wells are spaced from about 150 feet apart for Silverado wells to 850 feet apart for Lower San Pedro wells [Callison and Roth, 1967]. The early test wells were constructed using mild steel and drilled using cable tool, but the majority of the injection wells installed in the 1960s were asbestos cement casings

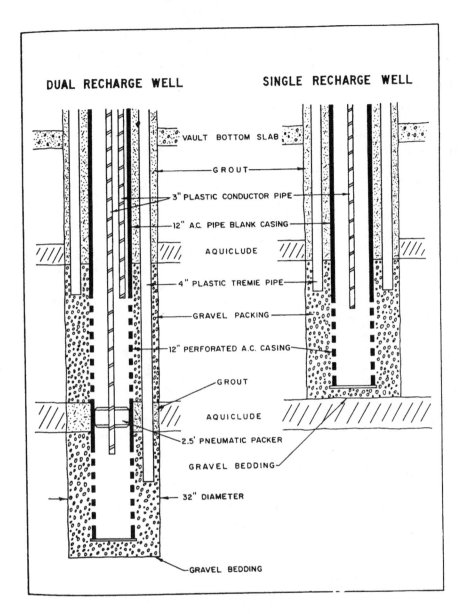

Figure 4: Typical injection well construction diagram.

surrounded by gravel envelopes drilled using reverse rotary. Newer wells have been constructed using stainless steel casing and screen.

Prior to 1995, the wells injected 100% potable water imported from northern California and the Colorado River. Since 1995, up to 50% of the injected water source has been from tertiary treated recycled water that also passes through reverse osmosis membranes. Plans are currently underway to increase the recycled water use up to 100% over the next 3 to 5 years. The use of recycled water will reduce or eliminate the need for the valuable imported water that will become available for direct potable use.

2.2   Alamitos Gap Barrier Project

The Alamitos Gap Barrier Project (ABP) was the second of the three barrier projects started. It was designed and constructed to protect the southeastern corner of the Central Basin and southwestern corner of the Orange County basin from saltwater intrusion. The ABP crosses the Los Angeles/Orange County line, and agencies from both counties participate in the operation and financing of the barrier.

The ABP originally consisted of 35 injection wells, 230 observation wells, and 4 extraction wells forming an approximate 2.2-mile long barrier arc across the Alamitos gap. The barrier is about 2 miles inland from the coast. About half of the wells were installed between 1965 and 1967 with the remaining original wells installed between 1977 and 1993. In 2000, nine additional injection wells and three observation wells were installed.

Pathways for saltwater intrusion are generally in the shallower aquifers, which have different nomenclature based on the adjacent Orange County groundwater basin. Injection occurs into the (from shallowest to deepest) C, B, A, and I aquifers (equivalent to the Lynwood aquifer [Lipshie and Larson, 1995]). The injection wells are completed to depths from about 100 feet below the ground surface to 450 feet, with most wells in the range of 200 to 450 feet in depth. The wells were constructed using Type 304 stainless steel casing [Johnson and Lundeen, 1967].

The ABP injects 100% potable water imported from northern California and the Colorado River. Starting in late 2002, the water source will be supplemented with up to 50% of tertiary treated recycled water that will pass through microfiltration, reverse osmosis membranes, and finally ultraviolet light for full treatment prior to injection to meet regulatory requirements.

2.3   Dominguez Gap Barrier Project

The Dominguez Gap Barrier Project (DGBP) was the last of the three barrier projects started. It was designed and constructed to protect the southern coastline of the West Coast Basin at the ports of Long Beach and Los Angeles from saltwater intrusion. The DGBP was started in 1969 and completed in 1971 with 41 injection wells and approximately 232 observation wells, extending 4.3 miles in length. In 2002, a total of 20 new injection wells at 10 locations were installed, and in 2003 a total of 33 more injection wells will be installed at 17 locations to improve portions of the barrier.

Pathways for saltwater intrusion are generally in the shallower aquifers (Gaspur, Gage, and Lynwood). However, inland from the barriers the shallower aquifers merge with the deeper aquifers, providing pathways to contaminate the heavily produced groundwater supply zones. The DGBP injection wells are completed into the Gage and Lynwood aquifers to try to stop the saltwater before it moves further inland. The injection wells are completed to depths from about 140 feet below the ground surface to 460 feet. The wells are spaced about 1,000 feet apart. The 20 new wells drilled in 2002 were placed in between existing wells to reduce the linear distance between wells and improve the mounding and protection effect of the pressure ridge. The original wells were constructed with asbestos cement casing, but the new wells have stainless steel casing.

The DGBP injects 100% potable water imported from northern California and the Colorado River. Starting in 2002, the water source will be supplemented with up to 50% of tertiary treated recycled water that will pass through reverse osmosis membranes for final treatment.

3. BARRIER WELL MANAGEMENT

Management of the saltwater barriers is an around the year task and is accomplished by multiple agencies with individual and shared responsibilities. The LACFCD owns, operates, and maintains the three barrier well projects and performs regular monitoring and sampling of the wells. Barrier effectiveness is measured by monitoring chloride concentrations and theoretical protective elevations using the Ghyben-Herzberg principal at internodal observation wells. Injection pressures are adjusted to maintain saltwater intrusion protection. A telemetry system is currently being installed or designed for all three barrier well projects to automate and increase data collection activities.

The Orange County Water District (OCWD) pays for the water, operations, and maintenance costs for the portion of the ABP that is within Orange County. The California Department of Health Services and Regional Water Quality Control Boards issue permits for the recycled water injection and ensure the water meets certain quality standards. All agencies involved with the barrier systems work together to ensure continued protection against saltwater intrusion while attempting to minimize costs.

The WRD purchases the water that is injected into the WCBBP, DGBP, and that portion of the ABP that is within Los Angeles County. WRD also teams with other agencies, including the USGS, the LACFCD, the OCWD, the United States Bureau of Reclamation (USBR), and consulting firms to investigate methods to improve the efficiency of the barrier well systems. For example, the WRD, the LACFCD, and the USGS are partnered to investigate the dynamics of saltwater intrusion through a monitoring and modeling program that has resulted in tools to determine various groundwater management options for the basins. The complexities of saltwater intrusion pathways and presence of saltwater sources in the aquifers were investigated through continuous coring boreholes to over 1,000 feet in depth, installation of aquifer specific, multi-level monitoring wells, geochemical sampling, geophysical logging, and flow-meter surveys.

The extensive amounts of data were incorporated into a Geographic Information System using Arc/Info software. A quasi three-dimensional groundwater flow model was constructed using Modflow to simulate the onshore/offshore aquifer systems and to simulate groundwater basin flow characteristics in response to pumping and recharge under historical and future scenarios [Reichard *et al.,* 2002]. Model optimization runs were performed to determine how to minimize barrier water injection costs while maintaining protection against saltwater intrusion. Specific inland extraction wells were identified as being the most likely candidates for contributing to the continued saltwater intrusion problem.

Modeling optimization results indicated that injection costs can be significantly reduced by raising groundwater elevations in the interior of the basins through reduced pumping (in-lieu delivery of surface water) and by construction of new spreading facilities. The unit costs for these alternatives are $219/acre-foot and $303/acre-foot, respectively, compared to the unit cost of non-interruptible imported barrier water at $528/acre-foot [Johnson *et al.*, 2001].

4. BARRIER WELLS ALTERNATIVES

The barrier well injection water costs have increased 2,445% since 1960, from $20.75/acre-foot to a maximum of $528/acre-foot. As shown previously in Table 1, in water year 2000/2001 a total of 30,382 acre-feet of water was injected into the barriers at a cost of nearly $15,000,000. The injection water is purchased by the Water Replenishment District of Southern California, who raises the money by placing a replenishment assessment on each acre-foot of groundwater pumped within the CWCB. The residents and businesses overlying the basins ultimately pay for the cost of this water. With plans for additional injection wells in the future and forecasted rising water rates, the injection water costs could nearly double over the next 5 years.

To address the high cost of injection water, the WRD, the LACFCD, and the USBR commissioned a feasibility study to find viable alternatives to injection wells [URS Greiner, Woodward-Clyde, 1999]. The Santa Ana Office (California) of URS Greiner Woodward Clyde conducted the study. A copy of the report is contained in the accompanying CD.

Nine types of alternative seawater barriers were identified, including slurry walls, deep soil mixing, grout curtains, jet grouting, *in situ* vitrification, channel lining, rubber dams, air (nitrogen gas) injection, and biological barrier walls. A description of each of these potential barrier alternatives is described below:

Slurry Wall: Construction of a slurry wall involves the excavation of a bentonite slurry stabilized trench and the final displacement filling of the trench with soil-bentonite, cement-bentonite, or plastic concrete materials that forms an impervious barrier to water flow. Depths up to 400 feet have been achieved. Because of the need to excavate a deep trench, there may be limitations due to near surface utilities.

Deep Soil Mixing: The DSM system utilizes a crane supported set of leads which guide a series of 36-inch diameter overlapping mixing shafts or augers that drill vertically into the soil while injecting a cement-based grout or bentonite. By combining the auger flights and mixing blades, the soil is lifted and blended with the grout in a pugmill fashion to create a series of overlapping soilcrete or soil-bentonite columns. By overlapping adjacent sets of these columns, a continuous cutoff wall is created to depths up to 100 feet.

Grout Curtains: Construction of a grout curtain involves drilling 2- to 6-inch diameter holes along a single line or multiple parallel lines. Grout is injected into the holes under pressure to fill the surrounding soil pores. By placing these holes in a tight enough spacing, a grout barrier is created. They can be installed to most any depth and can be surgically injected to treat specific zones. Materials other than grout can also be used, such as chemical grouts, microfine-cements, or clay slurries.

Jet Grouting: This method is an *in situ* mixing method commonly used to treat zones of soil in areas that are difficult to access. A small drill rig is used to advance a 2- to 3-inch diameter drill rod into the soil. A large pump is connected and used to pump grout into the soil at pressures of 4,000 to 6,000 pounds per square inch (psi). Horizontal ports in the drill rod direct the high-pressure grout flow (jet) into the soil. The rod is rotated and slowly raised, creating a cylindrical column of mixed soil and grout. By overlapping these individual grout columns, a continuous barrier is constructed. It presently has a depth limitation of about 100 feet.

***In Situ* Vitrification:** This technology was originally developed to contain hazardous materials within the soil. *In situ* vitrification uses specialized equipment to heat the soils to their fusion temperature (approximately 3,000 degrees Fahrenheit) utilizing electrical resistance heating. Organic components in the soil are pyrolyzed while the remaining soil particles are fused into a glass-like structure of melted silicate compounds. It has generally been applied above the water table.

Channel Liners: In the areas were saltwater intrusion was occurring by percolation through the unlined saltwater channels near the coast, a concrete channel liner could be constructed to impede the vertical movement of the water.

Rubber Dams: Similarly to the channel liners, inflatable rubber dams could be constructed in the channels to both prevent the landward movement of the saltwater further inland, and also to pond up freshwater on the upstream side of the dam. This would cause freshwater to percolate into the aquifers instead of the saltwater. These types of rubber dams are commonly used in Los Angeles County to cause groundwater recharge in sandy-bottom river channels.

Air (or Nitrogen Gas) Injection: Air injection is used in the development of oil and gas fields during tunneling to cut off the flow of water. This technology is largely untested for use in saltwater control. However, it is known that compressed air injected into the groundwater will cause a piezometric rise in water levels that can be used to alter groundwater gradients and flow directions. Air entrained in soil pores also causes an overall decrease in the permeability of the aquifers, which can be used to reduce flow. Modeling performed in the URS Greiner report [1999] showed that nitrogen could be used as a more stable gas and that the created pressure head would impede saltwater intrusion.

Biological Wall: This technology reports to use microbial biobarriers to manipulate the permeability and mass transport properties of the aquifer matrix. It essentially provides nutrients to injected, starved bacteria cultures so they grow and develop a reactive subsurface bio-barrier or bio-wall. The colonies grow and plug up the pore spaces of the aquifer, thereby creating a barrier to saltwater intrusion. The wall is maintained by the periodic injection of additional nutrients to feed the bacteria. This type of wall has been tested in the laboratory and field scale.

Two problem areas at each of the three barrier projects where saltwater intrusion is still occurring were identified during the study. The identified alternatives along with the traditional water injection and water extraction technologies were subjected to a thorough economic and technical decision making process to derive a recommended technology for each of the six identified barrier improvement areas. New barrier alternatives were recommended as the primary option over traditional injection wells at four of the six improvement areas. Pilot testing of the nitrogen gas injection alternative is currently underway, and preliminary designs for deep soil mixing are completed.

4.1 Decision Analysis

The economic and technical decision making process used for ranking of alternatives is commonly called decision analysis. It is a formal method to make decisions for problems characterized by multiple and often conflicting objectives, many stakeholders, and uncertainties in assessing the impacts of alternatives. The method divides the overall problem into smaller components, each of which can be analyzed more effectively, and then integrates the results of each component using the principals of rational behavior. In practice the steps are iterative. That is, an earlier step may need to be reviewed and revised in light of the assessments made in a subsequent step. The steps are summarized below.

Step 1. Identify alternatives. Each identified saltwater barrier alternative is defined in sufficient detail to allow the assessment of its impact on the objectives of interest. The details should include such items as areal boundaries; frequency and duration of application; and technology, equipment, and procedures to be used. For comparison, the existing water injection barrier is included in the evaluated alternatives.

Step 2. Define goals and measures. Goals are defined (what are we trying to achieve?) and one or more measures (how well is the goal achieved?) are defined for each goal. Goals for an alternative seawater barrier include: maximize system reliability, reduce implementation time, minimize environmental impact, and minimize cost. Examples of measures for the maximize system reliability goal are: years of experience with a given technology, sensitivity of system performance to subsurface conditions, system efficiency and effectiveness as a saltwater intrusion barrier, and ability of the system to withstand natural hazards (such as earthquakes).

Step 3. Assess preferences. The goal of this step is to assess the preferences of decision makers regarding the selected goals and measures.   The preferences are used to define a value that provides the means to assess the overall value of an alternative taking into account the impact of the alternative on the selected measures.

The preferences are assessed in two parts. First, preferences are assessed for each measure. These preferences define individual value functions that are used to convert the levels of individual measures into common value units.   Second, value tradeoffs are assessed between conflicting measures. The value tradeoffs define the relative weights of different measures that are used to combine the individual value of all measures into an overall value.

Preferences are value judgments that vary from one person to another. Decision analysis provides a structure for assessing these value judgments through structured interviews with a representative group of decision makers and their delegates. An explicit statement of subjective value judgments made by individuals provides a better understanding of individual perspectives. This process helps to define common ground, identify differences of opinion, and develop a reasonable consensus relative to key value judgments.

Step 4. Estimate the impacts of alternatives. The goal of this step is to estimate the impacts of each alternative in terms of the selected measures. The results are summarized in an impact matrix in which rows represent alternatives and columns represent measures. Each cell in the matrix is an impact estimate of a given alternative on a specified measure. If the impact on a given measure is highly uncertain, it can be analyzed by defining a range of possible impact and assessing the probabilities of impact.

Step 5. Evaluate and rank alternatives. This step integrates the information from the previous steps and an overall value is computed for each alternative. The alternatives are then ranked in descending order of overall value.

4.2   Decision Analysis Results

Sensitivity evaluation of the ranking to the various assumptions and value judgments is an important part of decision analysis. As an example, the acceptable value tradeoffs between conflicting measures may vary among stakeholders. The degree of acceptability of various alternatives among stakeholders can be evaluated by examining the influence of the different value tradeoffs on the overall value of each alternative. Sensitivity analysis results assist in identifying one or more alternatives that are consistently ranked high under a variety of reasonable value judgments, and are therefore likely to be widely accepted.

The decision analysis rated several alternative saltwater barrier types higher than the existing water injection at several specific locations. The analysis suggested a passive deep soil mixed (DSM) wall at the western end of the Dominguez Gap Barrier and at an alternate location to the Alamitos Barrier where the aquifer becomes shallower. The existing water injection barrier was the preferred barrier type at the north end of the West Coast Basin Barrier. An air/nitrogen injection barrier, contingent on the success of a pilot field test, was the preferred barrier at the deeper south end of the West Coast Basin Barrier. Table 2 summarizes the results.

| Description | Cost ($M) | Ranking[1] | |
		Pre-Cost[2]	Post-Cost[2]
Alamitos Area 1			
Water Injection	27	1	3
Deep Soil Mixing	7	2	1
Slurry Wall	12	2	2
Water Extraction	52	3	5
Biological Wall	18	4	4
Alamitos Area 2			
Rubber Dam	3	1	1
Channel Lining 1	8	2	2
Channel Lining 2	4	3	3
Dominguez Area 1			
Water Injection	11	1	1
Slurry Wall	24	2	2
Grout Curtain	20	3	3
Biological Wall	20	4	5
Water Extraction	13	5	4
Dominguez Area 2			
Slurry Wall	8	1	2
Deep Soil Mixing	4	2	1
Water Injection	6	3	3
Jet Grouting	11	4	4
Biological Wall	8	5	5
Water Extraction	13	6	6
West Coast Area 1			
Slurry Wall	42	1	5
Water Injection	10	2	1
Nitrogen Gas Wall	5	3	2
Grout Curtain	20	4	4
Biological Wall	12	5	3
West Coast Area 2			
Water Injection	20	1	2
Nitrogen Gas Wall	5	2	1
Grout Curtain	30	3	4
Biological Wall	20	4	3

(1) 1 = Best; 6 = Worst

(2) Pre-Cost rating does not consider cost. Post-Cost rating includes cost.

Table 2: Decision analysis results.

5. CONCLUSIONS

Injection wells have been successfully used to both control saltwater intrusion and to replenish the overdrafted Central and West Coast basin aquifers of Los Angeles County since the early 1950s. However, the rising cost of the injection water requires alternatives to injection wells to be explored. Computer modeling was performed to optimize groundwater pumping patterns in the basins to maximize groundwater production while minimizing the drop in groundwater elevations at the coast to minimize the injection requirements at the saltwater barrier wells.

A feasibility study was also performed to identify alternatives to saltwater intrusion control other than from the use of injection wells. The study addressed the potential physical alternatives and used a decision analysis approach to test each alternative at six locations along the current barrier alignment.

Based on the results, it was concluded that three barrier alternatives appear more cost effective than injection wells at certain locations, including deep soil mixing at the DGBP-Area 2 and the ABP-Area 1. Nitrogen gas injection was judged the best alternative at the WCBBP-Area 2. The rubber dam was the best alternative at the ABP-Area 1. At the other barrier problem areas, continued use of injection wells for intrusion control proved to be the best alternative.

A field test is currently underway to evaluate the nitrogen gas injection alternative at the WCBBP. An existing barrier injection well has been identified and will be equipped to have nitrogen gas injected into it for a period of time while water levels in nearby monitoring wells are measured to determine if water levels rise. An important criterion for the test to work is a competent aquitard overlying the injected aquifer to trap the buildup of pressure as the gas is injected.

The deep soil-mixing alternative is in the planning stages for a field test by the Water Replenishment District of Southern California, the Los Angeles County Department of Public Works, and the U.S. Bureau of Reclamation. The test will involve constructing about a 500-foot reach of the wall and testing the pre- and post-wall hydraulic effects across the barrier. If the test is successful, a full-scale deep soil mixing wall may be constructed.

Acknowledgments

The following individuals provided valuable contributions for this chapter: Dr. Eric Reichard of the United States Geological Survey, Wayne Jackson of the Los Angeles County Department of Public Works, Dennis Watt of the United States Bureau of Reclamation, Stephen Thomas of Camp

Dresser & McKee, Inc., and John Barneich of URS Greiner, Woodward Clyde (currently with GeoPentech Incorporated). Wanjiru Njuguna of the Water Replenishment District of Southern California prepared graphics for this paper.

REFERENCES

California Department of Water Resources, Planned Utilization of the Ground Water Basins of the Coastal Plain of Los Angeles County, Appendix A—Ground Water Geology, Bulletin No. 104, 1961.

California Department of Water Resources, Planned Utilization of the Ground Water Basins of the Coastal Plain of Los Angeles County, Appendix B—Safe Yield Determinations, Bulletin No. 104, 1962.

California Department of Water Resources, Seawater Intrusion in California. Appendix B by the Los Angeles County Flood Control District, Bulletin No. 63, 1957.

California Department of Water Resources, Sea Water Intrusion into Ground Water Basins Bordering the California Coast and Inland Bays, Water Quality Investigation Report No. 1, 1950.

Callison, J.C. and Roth, J.N., "Construction geology of the west coast basin barrier project," *Engineering Geology*, 4(2), 1967.

Johnson, M. and Lundeen, E.W, "Alamitos barrier project – Resume of geohydrologic investigation and status of barrier construction," *Engineering Geology*, 4(1), 1967.

Johnson, T., Reichard, E., Land, M., and Crawford, S., "Monitoring, modeling and managing saltwater intrusion, Central and West Coast Basins, Los Angeles County, California," CDROM Proceedings, the First International Conference on Saltwater Intrusion and Coastal Aquifers, D. Ouazar and A.H.-D. Cheng, Eds., Essaouira, Morocco, April 2001.

Lipshie, S.R. and Larson, R.A., "The West Coast Basin, Dominguez Gap, and Alamitos Seawater-Intrusion Barrier System, Los Angeles and Orange Counties, California," *AEG News*, 38(4), 25-2, 1995.

Mendenhall, W.C., Development of Underground Waters in the Eastern Coastal Plain Region of Southern California, United States Geological Survey Water-Supply Paper 137, 1905.

Poland, J.F., Garrett, A.A., and Sinnott, A., Geology, Hydrology, and Chemical Character of the Ground Waters in the Torrance-Santa Monica Area, Los Angeles County, California, U.S.G.S. Water Supply Paper 1461, 1959.

Reichard, E., Land, M., Crawford, S., Schipke Paybins, K., Nishikawa, T., Everett, R., and Johnson, T., Geohydrology, Geochemistry, and Ground-Water Simulation-optimization of the Central and West Coast Basins, Los Angeles County, California. United States Geological Survey Water Resources Investigation Report 02-xxxx (unassigned at this time). Prepared in co-operation with the Water Replenishment District of Southern California, 2002.

URS Greiner, Woodward-Clyde, Final Report: Alternative Seawater Barrier Feasibility Study, 1999.

Water Replenishment District of Southern California, Engineering Survey and Report, 2001.

Water Replenishment District of Southern California, Engineering Survey and Report, 2002.

CHAPTER 3

MODFLOW-Based Tools for Simulation of Variable-Density Groundwater Flow

C.D. Langevin, G.H.P. Oude Essink, S. Panday, M. Bakker,

H. Prommer, E.D. Swain, W. Jones, M. Beach, M. Barcelo

1. INTRODUCTION

Most scientists and engineers refer to MODFLOW [McDonald and Harbaugh, 1988; Harbaugh and McDonald, 1996; Harbaugh *et al.*, 2000] as the computer program most widely used for constant-density groundwater flow problems. The success of MODFLOW is largely attributed to its thorough documentation, modular structure, which makes the program easy to modify and enhance, and the public availability of the software and source code. MODFLOW has been referred to as a "community model," because of the large number of packages and utilities developed for the program [Hill *et al.*, 2003]. In recent years, the MODFLOW code has been adapted to simulate variable-density groundwater flow. Because MODFLOW is so widely used, these variable-density versions of the code are rapidly gaining acceptance by the modeling community.

To represent variable-density flow in MODFLOW, the flow equation is formulated in terms of equivalent freshwater head. With this approach, the finite-difference representation is rewritten so that fluid density is isolated into mathematical terms that are identical in form to source and sink terms. These "pseudo-sources" can then be easily incorporated into the matrix equations solved by MODFLOW. Weiss [1982] was one of the first to recast the groundwater flow equation in terms of equivalent freshwater head and introduce the concept of a pseudo-source. Lebbe [1983] used a similar approach to develop a variable-density version of the MOC code [Konikow and Bredehoeft, 1978]. Maas and Emke [1988] were among the first to incorporate variable-density flow into MODFLOW. The approach was improved by Olsthoorn [1996] to account for inclined model layers. These initial studies allowed for fluid density to vary in space, but not in time. Recently, solute transport codes have been linked directly with MODFLOW to represent the transient effects of an advecting and dispersing solute

1-56670-605-X/04/$0.00+$1.50
© 2004 by CRC Press LLC

concentration field on variable-density groundwater flow patterns. These MODFLOW-based codes are being applied to numerous hydrologic problems involving variable-density groundwater flow.

Descriptions and applications of four of the commonly used MODFLOW-based computer codes are presented in this chapter. The four codes (SEAWAT, MOCDENS3D, MODHMS, and the Sea Water Intrusion Package for MODFLOW-2000) have been applied to case studies and have been documented and tested with variable-density benchmark problems. The first three programs represent advective and dispersive solute transport. The fourth program uses a non-dispersive, continuity of flow approach to simulate movement of multiple density isosurfaces.

2. SEAWAT

C.D. Langevin, H. Prommer, E.D. Swain

The SEAWAT computer program is designed to simulate a wide range of hydrogeologic problems involving variable-density groundwater flow and solute transport. The SEAWAT code has been applied worldwide to evaluate such problems as saltwater intrusion, submarine groundwater discharge, aquifer storage and recovery, brine migration, and coastal wetland hydrology. The source code, documentation, and executable computer program are available to the public at the USGS web page.[1]

This section provides a brief description of the SEAWAT program and presents applications of SEAWAT to geochemical modeling and integrated surface water and groundwater modeling. Additional information, including the SEAWAT documentation, is available on the accompanying CD.

2.1 Program Description

SEAWAT was designed by combining MODFLOW-88 and MT3DMS into a single program that solves the coupled variable-density groundwater flow and solute-transport equations [Guo and Bennett, 1998; Guo and Langevin, 2002]. The flow and transport equations are coupled in two ways. First, the fluid velocities that result from solving the flow equation are used in the advective term of the solute-transport equation. Second, the solute-transport equation is solved, and an equation of state is used to calculate fluid densities from the updated solute concentrations. These fluid densities are then used directly in the next solution to the variable-density groundwater flow equation.

---

[1] http://water.usgs.gov/ogw/seawat/

The variable-density groundwater flow equation solved by SEAWAT is formulated using equivalent freshwater head as the principal dependent variable. In this form, the equation is similar to the constant-density groundwater flow equation solved by MODFLOW. Thus, with minor modifications, MODFLOW routines are used to represent variable density groundwater flow. Modifications include conservation of fluid mass, rather than fluid volume, and the addition of relative density difference terms, or pseudo-sources. The procedure for solving the variable-density flow equation is identical to the procedure implemented in MODFLOW. Matrix equations are formulated for each iteration, and a solver approximates the solution. Modifications are not required for the MT3DMS routines that solve the transport equation.

Like MT3DMS, SEAWAT divides simulations into stress periods, flow timesteps, and transport timesteps. The lengths for stress periods and flow timesteps are specified by the user; however, the time lengths for transport timesteps are calculated by the program based on stability criteria for an accurate solution to the transport equation. Because flow and transport are coupled in SEAWAT, either explicitly or implicitly, the flow and transport equations are solved for each transport timestep. This requirement does not apply for simulations with standard MODFLOW and MT3DMS because, in those cases, concentrations do not affect the flow field.

Output from SEAWAT consists of equivalent freshwater heads, cell-by-cell fluid fluxes, solute concentrations, and mass balance information. This output is in standard MODFLOW and MT3DMS format, and most publicly and commercially available software can be used to process simulation results. For example, animations of velocity vectors and solute concentrations can be prepared using the U.S. Geological Survey's Model Viewer program [Hsieh and Winston, 2002], and post-processing programs such as MODPATH [Pollock, 1994] can be used to perform particle tracking using SEAWAT output.

The U.S. Geological Survey actively supports the SEAWAT program. As new packages, processes, and utilities are added to the MODFLOW and MT3DMS programs, these improvements are incorporated into SEAWAT. For example, a new version of SEAWAT, which is based on MODFLOW-2000, was recently developed.

2.2   Reactive Transport Modeling with PHREEQC and SEAWAT

Two disciplines, namely, reactive transport modeling and variable-density flow modeling, have received significant attention over the past two decades. Well-known representatives of the former class of models are, for example, MIN3P [Mayer *et al.*, 2002], GIMRT/CRUNCH [Steefel, 2001],

PHREEQC [Parkhurst and Appelo, 1999], PHAST [Parkhurst *et al.*, 1995], HydroBioGeoChem [Yeh *et al.*, 1998], and some MODFLOW/MT3DMS-based models such as RT3D [Clement, 1997] and PHT3D [Prommer *et al.*, 2003].

    In most cases the separation of the two disciplines is well justified, because (i) density gradients are small enough to be of negligible influence on the reactive transport of multiple solutes or (ii) reactions, in particular water-sediment interactions such as mineral dissolution/precipitation and/or sorption, have a minor effect on the density of the aqueous phase. However, specific cases exist where transport phenomena can only be accurately described by considering simultaneously both variable density and reactive processes. For example, Zhang *et al.* [1998] were only able to explain the differential downward movement of a lithium (Li^+) and a bromide (Br^-) plume at Cape Cod through multi-species transport simulations that considered the variable density of the plume(s) and lithium sorption. Furthermore, Christensen *et al.* [2001, 2002] demonstrated the interactions between reactive processes and density variations for (i) a controlled seawater intrusion experiment, where seawater was forced inland by pumping, thereby undergoing reactions such as Na/Ca exchange, calcite dissolution-precipitation, sulfate-reduction, and FeS precipitation, and (ii) for a landfill leachate plume, where the density influences the distribution of the redox-species and buffering reactions by Fe and Mn hydroxides. The ongoing project to combine SEAWAT with the geochemical model PHREEQC-2 was initially motivated by the desire to simulate and quantify reactive changes that occur as a result of tidally induced, variable density flow near the aquifer/ocean interface.

    The governing equation for both transport and reactions of the ith (mobile) aqueous species/component, solved by the coupled model, is:

$$\frac{\partial C_i}{\partial t} = \frac{\partial}{\partial x_\alpha}\left(D_{\alpha\beta} \frac{\partial C_i}{\partial x_\beta} \right) - \frac{\partial}{\partial x_\alpha}\left(v_\alpha C_i \right) + r_{reac,i} \qquad (1)$$

where v_α is the pore-water velocity in direction x_α, $D_{\alpha\beta}$ is the hydrodynamic dispersion coefficient tensor, and $r_{reac,i}$ is a source/sink rate due to the chemical reactions that involve the ith aqueous component. C_i is the total aqueous component concentration [Yeh and Tripathi, 1989], defined as:

$$C_i = c_i + \sum_{j=1,n_s} Y_j^s s_j \ , \qquad (2)$$

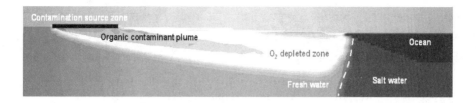

Figure 1: Simulated coastal point source pollution by an aerobically degrading organic contaminant.

where c_i is the molar concentration of the (uncomplexed) aqueous component, n_s is the number of species in dissolved form that have complexed with the aqueous component, Y_j^s is the stoichiometric coefficient of the aqueous component in the jth complexed species, and s_j is the molar concentration of the jth complexed species. As in PHT3D, the (local) redox-state, pe, is modeled by transporting chemicals/components in different redox states separately, while the pH is modeled from the (local) charge balance.

Coupling of PHREEQC-2 with SEAWAT is achieved through a sequential operator splitting technique [Yeh and Tripathi, 1989; Barry *et al.*, 2002], similar to the technique used for the PHT3D model, which couples PHREEQC-2 with MT3DMS. The splitting scheme used to solve the advection-dispersion-reaction equation (Eq. (1)) for a user-defined time step length consists of two steps. In the first step the advection and dispersion term of mobile species/components is solved with SEAWAT for the time step length Δt. In the subsequent step the reaction term r_{reac} in Eq. (1) is solved through grid-cell wise batch-type PHREEQC-2 reaction calculations. This step accounts for the concentration changes that have occurred during Δt as a result of reactive processes. The reaction term r_{reac} in Eq. (1) corresponds to the computed concentration differences from before (PHREEQC-2 input concentrations) and after the reaction step (PHREEQC-2 output concentrations).

Figure 1 illustrates the results from one of the initial (simple) multi-species test simulations of coastal point-source pollution by an organic contaminant. The plume is degraded aerobically, i.e., the degradation reaction creates an oxygen-depleted zone in an aquifer containing groundwater of variable density.

2.3   Integrated Surface-Water and Groundwater Modeling with SWIFT2D and SEAWAT

2.3.1   Code Description

To simulate the coastal hydrology of the southern Everglades of Florida, which is characterized by shallow overland flow and subsurface groundwater flow, SEAWAT was coupled with the hydrodynamic estuary model, SWIFT2D (Surface-Water Integrated Flow and Transport in 2-Dimensions) [Langevin et al, 2002; Langevin et al., 2003; Swain et al., 2003]. SWIFT2D solves the full dynamic wave equations, including density effects, and can also represent transport of multiple constituents, such as the dissolved species in seawater. The SWIFT2D code was originally developed in the Netherlands [Leendertse, 1987], and was later modified by the U.S. Geological Survey to represent overland flow in wetlands by including spatially varying rainfall, evapotranspiration, and wind sheltering coefficients [Swain et al., 2003].

The coupling of SWIFT2D and SEAWAT is accomplished by including the programs as subroutines of a main program called FTLOADDS (Flow and Transport in a Linked Overland-Aquifer Density Dependent System). FTLOADDS uses a mass conservative approach to couple the surface water and groundwater systems, and computes leakage between the wetland and the aquifer using a variable-density form of Darcy's Law written in terms of equivalent freshwater head. The leakage representation also includes associated solute transfer, based on leakage rates, flow direction, and solute concentrations in the wetland and aquifer.

Coupling between SWIFT2D and SEAWAT occurs at intervals equal to the stress period length in the groundwater model. For each stress period, which is one day in the current Everglades application, SWIFT2D is called first, using short timesteps, such as 15 minutes, to complete the entire groundwater model stress period. Within the SWIFT2D subroutine, leakage is calculated as a function of the surface water stage and the groundwater head from the end of the previous stress period. The total leakage volumes (for each cell) are summed for the stress period by accumulating the product of the leakage rate and the length of the surface water timestep. After SWIFT2D completes the stress period, the total leakage volumes are applied on a cell-by-cell basis to SEAWAT as it runs for the same stress period to calculate groundwater heads and solute concentrations.

FTLOADDS also accounts for the net solute flux between surface water and groundwater. When the leakage volume is computed for a surface-water timestep, the solute flux is computed based on flow direction. If the flow is upward from the aquifer into the wetland, the solute flux is calculated

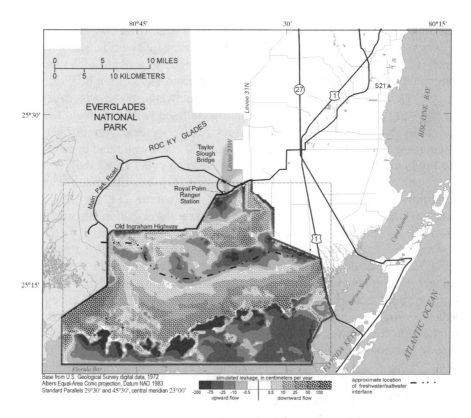

Figure 2: Map of southern Florida showing SICS model domain and simulated values of average daily leakage between surface water and groundwater.

by multiplying leakage volume and groundwater salinity. The calculated solute mass is then added to the surface-water cell in the SWIFT2D transport subroutine. If flow is downward from the wetland into the aquifer, the solute mass flux is calculated as the product of leakage volume and surface-water salinity. The total solute mass flux is summed for the surface-water timesteps and divided by the total leakage volume. This gives an equivalent salinity concentration for the total leakage over the stress period. Whichever direction of the leakage, the computed equivalent salinity is used in SEAWAT as the concentration of the water added or removed from the aquifer as leakage.

2.3.2   Application to the Southern Everglades of Florida

As part of the Comprehensive Everglades Restoration Plan, the U.S. Geological Survey has applied the FTLOADDS model to the Taylor Slough area in the southern Everglades of Florida (Figure 2) [Langevin *et al.*, 2002].

The finite-difference grid consists of 148 columns and 98 rows. Each cell is square with 304.8 m per side. The three-dimensional grid has 10 layers (each 3.2 m thick) and extends from land surface to a depth of 32 m. The integrated model simulates flow and transport from 1995 through 1999.

The integrated surface water and groundwater model was calibrated by adjusting model input parameters until simulated values of stage, salinity, and flow matched with observed values at the wetland and Florida Bay monitoring sites. Daily leakage rates between surface water and groundwater are produced as part of the model output for each cell. These daily leakage rates were averaged over the 5-year simulation period to illustrate the spatial variability in surface water/groundwater interaction (Figure 2). These leakage rates do not include recharge or evapotranspiration directly to or from the water table. The model suggests an alternating pattern of downward and upward leakage from north to south (Figure 2). To the north, most leakage is downward into the aquifer, except near the Royal Palm Ranger station where upward flow occurs near Old Ingraham Highway. Further south, a large area of upward leakage exists. This area of upward leakage roughly corresponds with the location of the freshwater/saltwater interface in the aquifer. In this area, groundwater flowing toward the south moves upward where it meets groundwater with higher salinity. To the south, leakage is downward into the aquifer. The Buttonwood Embankment, which is a narrow ridge along the Florida Bay coastline, separates the inland wetlands from Florida Bay. The embankment impedes surface water flowing south and increases wetland stage levels to elevations slightly higher than stage levels in Florida Bay. South of the Buttonwood Embankment, groundwater discharges upward into the coastal embayments of Florida Bay. This upward leakage in the model is caused by the higher water levels on the north side of the embankment. These model results suggest that surface water and groundwater interactions are an important component of the water budget for the Taylor Slough area.

3. MOCDENS3D

G.H.P. Oude Essink

3.1 Program Description

The computer code MOCDENS3D [Oude Essink, 1998, 2001] can simulate groundwater flow and coupled solute transport in porous media. The code is based on the United States Geological Survey public domain three-dimensional finite difference computer code MOC3D [Konikow *et al.*, 1996]. Density differences in groundwater are taken into account in the mathematical formulation. So-called freshwater heads and buoyancy term are introduced. As a result, it is possible to simulate non-stationary flow of fresh, brackish, and saline groundwater in coastal aquifers. More detail of the code is described in Oude Essink [1999]. Note that MOCDENS3D is similar to SEAWAT: the first uses MOC3D for solute transport, whereas the latter applies MT3DMS [Zheng and Wang, 1999].

3.2   Effect of Sea Level Rise and Land Subsidence in a Dutch Coastal Aquifer

3.2.1   Introduction to the Dutch Situation

Saltwater intrusion is threatening coastal groundwater systems in the Netherlands. At the root of the problem are both natural processes and anthropogenic activities that have been going on for centuries. Autonomous events, land subsidence, and sea level rise all influence the distribution of fresh, brackish, and saline groundwater in Dutch coastal aquifers.

The greatest land subsidence is occurring in the peaty and clayey regions in the west and north of the Netherlands and emanates from two, human-driven processes. The first—soil drainage—is a slow and continuous process that started about a thousand years ago when the Dutch began to drain their swampy land. The second—land reclamation—causes a relatively abrupt change in the surface level. In particular, it was the reclamation of the deep lakes during the past centuries that caused the strong flow of saline groundwater from the sea to the coastal aquifers. These so-called *deep polders* are currently experiencing upward seepage flow.

An example of a Dutch coastal aquifer will show that on the long term, the effects of sea level rise and land subsidence—in terms of the amount of seepage, average salt content, and salt load—can be considerable [Oude Essink and Schaars, 2003].

3.2.2   Model of the Groundwater System of Rijnland Water Board

The Rijnland Water Board has a surface area of about 1,100 km^2 (Figure 3a) and accommodates some 1.3 million people. Since the 12th century, the water board manages water quantity and water quality aspects in the area. Sand dunes are present at the western side of the water board (Figure 3b). Three major drinking water companies are active in the dunes: DZH (Drinking Water Company Zuid-Holland), GWA (Amsterdam Waterworks), and PWN (Water Company Noord-Holland).

Phreatic water levels in the dune areas can go up to more than 7 meters above mean sea level. At the inland side of the dune area, some large low-lying polder areas with controlled water levels occur (Figure 4a). The lowest phreatic water levels in the water board itself can be found northwest of the city Gouda (down to nearly −7 m N.A.P.) and in the Haarlemmermeer

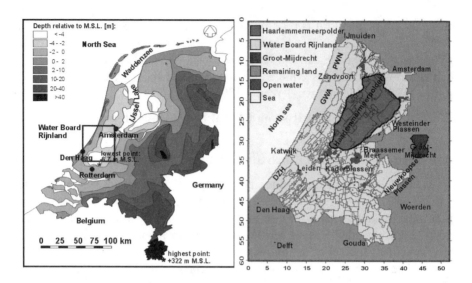

Figure 3: (a) Map of The Netherlands: position of the Rijnland Water Board and ground surface of the Netherlands; (b) Map of the Rijnland Water Board: position of some polder areas and the sand-dune areas of the drinking water companies DZH, GWA, and PWN. The Haarlemmermeer polder is also a part of the water board.

polder, where the airport Schiphol is located, with levels as low as −6.5 m N.A.P. Before the middle of the 19th century, a lake covered the Haarlemmermeer polder area. Due to flooding threats in the neighboring cities, this lake was reclaimed during the years 1840–1852 which caused a relative abrupt change in heads. Subsequently, a completely different groundwater flow regime was created regionally. In addition, the polder Groot-Mijdrecht, situated outside the water board, is also mentioned here. Though the surface area of this polder is not large, the phreatic water level is low (less than −6.5 m N.A.P.) and the Holocene aquitard on top of the groundwater system is very thin. Seepage in this area is very large (more than 5 mm/day) and groundwater from a large region around it is flowing to the polder at a rapid pace. Some large groundwater extractions from the lower aquifer system are taking place, up to 20 million m^3/yr at Hoogovens near IJmuiden.

The groundwater system consists of a three-dimensional grid of 52.25 km by 60.25 km (~3,150 km^2) by 190 m depth and is divided into a large number of elements. Each element is 250 m by 250 m in horizontal plane. In vertical direction the thickness of the elements varies from 5 m for the 10 upper layers to 10 m for the deepest 14 layers (Figure 4b). The grid

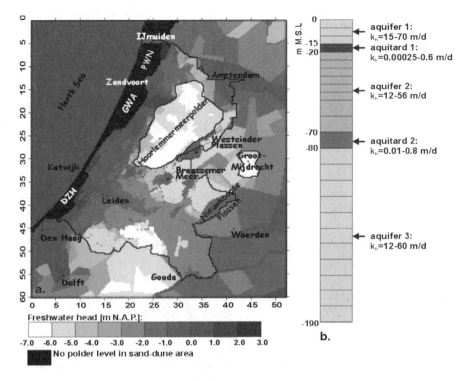

Figure 4: (a) Phreatic water levels or polder levels in the area (note that in the sand-dune areas, no polder levels are given); (b) Simplified subsoil composition of the bottom of the water board of Rijnland and hydraulic conductivity values.

contains 1,208,856 active elements: $n_x = 209$, $n_y = 241$, $n_z = 24$, where n_i denotes the number of elements in the i direction. Each element contains initially eight particles, which gives in total 9.6 million particles to solve the advection term of the solute transport equation. The flow time step Δt to recalculate the groundwater flow equation is 1 year. The convergence criterion for the groundwater flow equation (freshwater head) is equal to 10^{-4} m.

Data has been retrieved from NAGROM (The National Groundwater Model of The Netherlands). Figure 4b shows the composition of the groundwater system into three permeable aquifers, intersected by an aquitard in the upper part of the system and an aquitard of clayey and peat composite between −70 and −80 m N.A.P. For each subsystem, the interval of the horizontal hydraulic conductivity k_h is given in the figure. The anisotropy ratio k_z/k_x is assumed to be 0.1 for all layers. The effective porosity n_e is a bit low: 25%. The longitudinal dispersivity α_L is set equal to 1 m, while the ratio of transversal to longitudinal dispersivity is 0.1.

The bottom of the system is a no-flow boundary. Hydrostatic conditions occur at the four sides of the model. At the top of the system, the natural groundwater recharge in the sand-dune area varies from 0.94 to 1.14 mm/day. The water level at the sea is set to 0.0 m N.A.P. for the year 2000 AD. The general head boundary levels in the polder area are equal to the phreatic water level in the considered polder units, varying from +2.0 m near IJmuiden to −7.0 m N.A.P. northwest of Gouda.

At the initial situation (2000 AD), the hydrogeologic system contains saline, brackish as well as fresh groundwater. On the average, the salinity increases with depth, whereas freshwater lenses exist at the sand-dune areas at the western part of the water board, up to −90 m N.A.P. Freshwater from the sand dunes flows both to the sea and to the adjacent low-lying polder areas. The chloride concentration of the upper layers is already quite high in some low-lying polder areas such as the Haarlemmermeer polder and the polder Groot-Mijdrecht. The volumetric concentration expansion gradient β_C is 1.34×10^{-6} l/mg Cl⁻. Saline groundwater in the lower layers does not exceed 18,630 mg Cl⁻/l. The corresponding density of that saline groundwater equals 1,025 kg/m³.

Calibration was focused on freshwater heads in the hydrogeologic system, and to some extent on seepage and salt load values in the Haarlemmermeer polder and the polder Groot-Mijdrecht. Calibration data has been derived from the water board itself, the NAGROM database, ICW (1976), and the DINO database of Netherlands Institute of Applied Geosciences (TNO-NITG). The model was calibrated by comparing 1632 measured and computed freshwater heads, and for seepage and salt load values of some polders. Note that the measured heads are corrected for density differences. The mean error between measured and computed freshwater heads is −0.16 m, the mean absolute error 0.61 m, and the standard deviation 0.79 m.

3.2.3   Sea Level Rise and Land Subsidence

It is expected that climate change causes a rise in mean sea level and a change in natural groundwater recharge. As exact figures are not known yet, an average impact scenario is considered here by taking into account the most likely future developments in this area:

- According to the Intergovernmental Panel of Climate Change [IPCC, 2001], a sea level rise of 0.48 m is to be expected for the year 2100 (relative to 1990), with an uncertainty range from 0.09 to 0.88 m. Based on these figures, a sea level rise of 50 cm per century will be implemented at the North Sea, in steps of 0.005 m per time step of 1 year, from 2000 AD on.

- An instantaneous increase of natural groundwater recharge of 3% at all sand-dune areas in 2000 AD.

- Oxidation of peat, compaction and shrinkage of clay, and groundwater recovery are causing land subsidence, especially in the peat areas of the water board. The following values are inserted: a land subsidence of –0.010 m per year for the peat areas; no subsidence for the sand-dune areas; and –0.003 m per year for the rest of the land surface (respectively 25, 9, and 66% of the land surface in the entire modeled area).

- A reduction of groundwater extraction in the sand-dune areas GWA (–1.3 million m^3/yr) and PWN (–4.5 million m^3/yr).

The total simulation time is 200 years.

3.2.4   Discussion of Results

The overall picture is that the groundwater system will contain more saline groundwater these coming centuries. The numerical model supports the theory that the present situation is not in equilibrium from a salinity point of view. Figure 5 shows the chloride distribution at –2.5 and –7.5 m N.A.P. for the years 2000 and 2200 AD. Salinization is going on, especially in the areas close to the coastline. Though the differences look small due to the fact that groundwater flow and subsequently solute transport are slow processes, changes in seepage and salt load at the top aquifer system are pretty significant (Figure 6). The combination of autonomous development (reclamation of the deep lakes in the past), sea level rise, and land subsidence will intensify the salinization process: partly due to an increase of seepage values (+6% in 2050 and +12% in 2200, relative to now) but mainly due to the increase in salinity of the top aquifer system. As a result, the overall salt load in the water board is estimated to increase +38% in 2050 and even +79% in 2200, relative to now. The more rapid increase in salt load is caused by an increased salinization of the upper aquifers.

3.2.5   Conclusions

A model of the variable density groundwater flow system of the Rijnland Water Board is constructed to quantify the effect of past anthropogenic activities, climate change (rise in sea level and an increase in natural groundwater recharge in the sand-dune areas), and land subsidence in large parts of the area. The code MOCDENS3D is used to simulate density dependent groundwater flow under influence of the above mentioned stresses. Numerical computations indicate that a serious saltwater intrusion

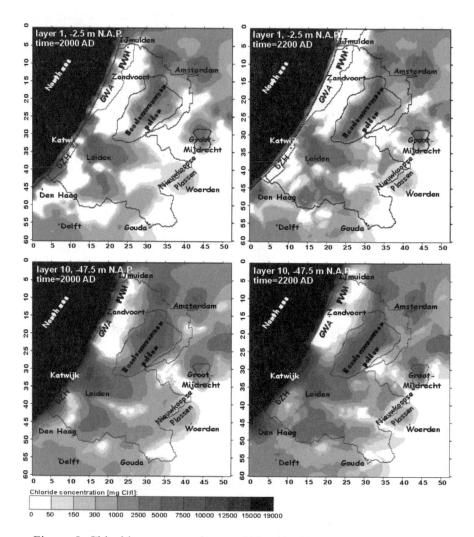

Figure 5: Chloride concentration at –2.5 and –47.5 m N.A.P. for the years
2000 and 2200 AD. Sea level rise and land subsidence is considered.

can be expected during the coming decennia, mainly because a large part of
the Rijnland Water Board is lying below mean sea level. The combined
effect for 2050 AD will be: a 6% increase of seepage and a 38% increase of
salt load in the Rijnland Water Board. The increase especially in salt load
will definitely affect surface water management aspects at the water board.

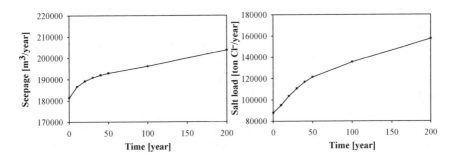

Figure 6: Seepage (in m³/day) and salt load (in ton Cl⁻/year) through the second model layer at −10 m N.A.P., summarized for the entire Rijnland Water Board, as a function of 200 years.

4. MODHMS

S. Panday, W. Jones, M. Beach, M. Barcelo

4.1   Program Description

4.1.1   Description

MODHMS [HydroGeoLogic, 2002] is a comprehensive hydrologic modeling system that extends MODFLOW to include the unsaturated zone, the overland flow domain, and channel/surface-water features. Contaminant transport routines are also incorporated for fate and transport calculations in single or dual porosity systems. Density coupling of the flow-field with concentrations of some or all species provides comprehensive analysis capabilities for complex coastal issues.

4.1.2   Physical Concepts and Model Features

MODFLOW's capabilities are expanded by MODHMS to solve the Richards equation for three-dimensional saturated-unsaturated subsurface flow, coupled with the diffusive wave equations for two-dimensional overland and one-dimensional channel flow (including effects of scale, pond storage, routing, and hydraulic structures). The primitive form of the transport equation for single or multiple species is also solved, with optional dual porosity considerations for the subsurface. Nonlinear adsorption and linear decay processes are incorporated, with provisions to accommodate user-supplied, complex reaction modules. Multi-phase transport occurs with equilibrium partitioning considerations and diffusion/storage/decay in the inactive (air) phase. Non-isothermal conditions may be simulated by allocating the temperature variable as the first species of solution. Density coupling (in surface and subsurface regimes) of flow with transport of some or all contaminant species is achieved via a linear density relationship with concentration (adjustment of viscosity and density for the conductance term may also be optionally applied). Fluid pressure is therefore affected by species concentration, and the advection and dispersion terms are affected by the resultant volumetric fluid fluxes. Various combinations of the above simulation capabilities may be used for optimal solution to a given problem.

In addition to MODFLOW's stress packages, MODHMS includes fracture-wells to handle multi-layer pumping and prevent overpumping; an unconfined recharge seepage-face package (for subsurface simulations only) for ponding and hill slope seepage issues; and a comprehensive evapotranspiration (ET) package that accounts for climatic conditions. Transport boundaries include mass fluxes at inflow nodes and prescribed concentrations anywhere in the domain. For density-dependent cases, the flow condition is optionally checked at every iteration or time-step at a constant head node, before applying a prescribed concentration condition.

4.1.3   Computational Aspects

The three-dimensional finite-difference grid of MODFLOW is used for subsurface discretization, with a corresponding two-dimensional grid for the overland domain, and a finite-volume discretization for the channel/surface-water body domain. Alternatively, an orthogonal curvilinear grid may be used for the overland and subsurface regimes. Surface/subsurface interactions are expressed fully implicitly, or via iterative/linked options. Newton-Raphson linearization may be used for the unsaturated or unconfined flow equations, and pseudo-soil functions (that are more robust for wetting/drying situations) may be used for unconfined systems where unsaturated effects are neglected. The transport equations are solved using mass conserved schemes with Total Variation Diminishing (TVD), upstream or midpoint spatial weighting, and implicit or Crank-Nicolson temporal weighting options. For density-dependent simulations, the flow equation is solved in terms of equivalent freshwater heads, and the density correction is applied via Picard iteration between the flow and transport equations. Adaptive time-stepping and under-relaxation formulas are based on all system non-linearities, for optimal speed and robustness. Solution options are provided for various combinations of transient and steady-state flow and transport analysis.

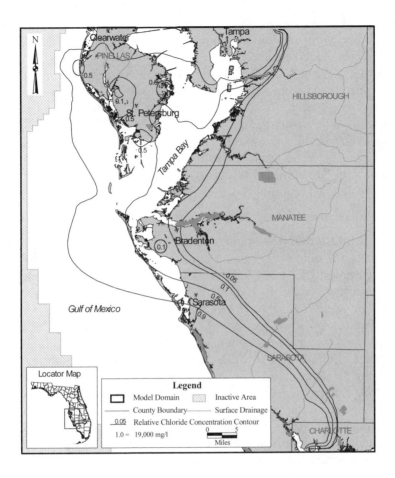

Figure 7: Location of study area and simulated chloride concentrations.

4.2 Case History

4.2.1 Site Location and Project Objectives

The study area lies in the southern portion of the Southwest Florida Water Management District (SWFWMD). The model domain includes all or portions of Pinellas, Hillsborough, Manatee, and Sarasota Counties and extends into the Gulf of Mexico covering approximately 60 miles by 100 miles (Figure 7). Management of saltwater intrusion due to significantly increased groundwater withdrawals was investigated using a density-dependent MODHMS model. Boundary conditions and model parameters were derived from a larger, regional MODFLOW model developed by the SWFWMD and referred to as the Southern District (SD) model. The local model was calibrated to available chloride and water level information from pre-development to current conditions. A steady-state pre-development calibration provided assumed hydrostatic equilibrium behavior of the flow/transport system under long-term average recharge conditions, and was followed by a post-development transient simulation using pumping estimates throughout the study area, from 1900 to 2000. The calibrated model was used to predict the impact of several potential water management scenarios from current conditions to 2050. Results of this analysis assisted in the development of a water level index that will aid in long-term management of groundwater resources. The CD accompanying this book contains the detailed report of this study [HydroGeoLogic, 2002].

4.2.2   Climate and Hydrogeologic Setting

The site location is humid and subtropical, characterized by warm wet summers and mild dry winters. Long-term rainfall averages 52 in/yr, and mean evapotranspiration is 39 in/yr. The underlying aquifers include the Surficial Aquifer System (SAS), the Intermediate Aquifer System (IAS), and the Floridan Aquifer System (FAS), each of which consists of permeable layers separated by lower permeability semi-confining units. The FAS is subdivided into major units comprising the Upper Floridan Aquifer (UFA), the Middle Confining Unit (MCU), and the Lower Floridan Aquifer (LFA), which is highly saline in this region and not a source of potable water. The UFA is the principal source of water in the region and is further subdivided into the Suwannee Limestone, the Ocala Limestone, and the Avon Park Formation, which consists of a main water-bearing zone overlying relatively lower-conductivity units. The MCU contains evaporites that are of extremely low conductivity and forms the bottom of the modeled system.

4.2.3   Conceptual Model and Calibration

The density-dependent saltwater intrusion model was developed from the SD model using telescoping mesh refinement, thereby maintaining the hydrostratigraphy, hydrogeologic properties, and imposed stresses of the regional model. Hydrogeologic units were further sub-divided vertically in the numerical grid of the local model to provide resolution for saltwater intrusion considerations. Only the FAS was considered for this study, therefore, recharge/discharge from the overlying IAS was obtained from regional flow model results and applied as a general head boundary across the overlying confining unit. The saltwater model domain included the Suwannee Limestone underlain by the Ocala limestone, the Avon Park Formation and the low conductivity Evaporite Zone of the MCU. Chloride and head conditions were prescribed underneath, for provision of upward

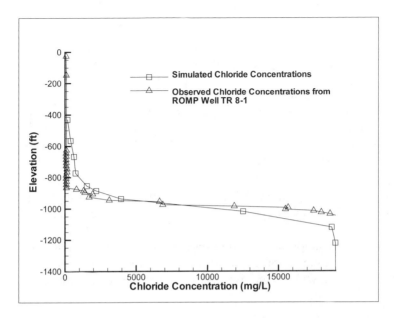

Figure 8: Comparison of observed and simulated chloride concentrations.

movement and upconing effects from deeper regions. Hydraulic conductivity values of the various formations were derived from the SD model transmissivity and leakance fields, and landward boundary conditions for each model run were obtained from parallel simulations using the SD model. Vertical anisotropies, dispersivities, and saltwater boundaries were obtained from field estimates or treated as calibration parameters. Conductivity fields were also adjusted slightly.

The model was first calibrated to steady-state environmental heads, and depth-to-chlorides (for 250, 500, and 1000 ppm levels) estimated for pre-development conditions (early 1900s), with further calibration for transients till the year 2000 (Figure 8). Calibration measures include collective statistics as well as temporal and depth-dependent heads and chloride concentrations obtained from individual wells. The model was used to predict the effects of different stresses within the SWUCA (400, 600, 800, and 1000 MGD) for the next 50 years, to determine relations between pumping, flow levels, and long-term saltwater intrusion in the FAS.

4.2.4   Model Calibration Results

The MODHMS model was able to accurately simulate hydraulic heads and chloride concentrations within the study area. The calibration to environmental heads was good, and the model adequately represented

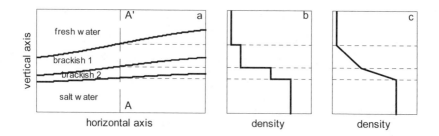

Figure 9: (a) Conceptual model with three surfaces, (b) density distribution of stratified flow, (c) density distribution of variable density flow.

depth-dependent chloride concentrations (Figure 8) and chloride movement from pre- to post-development conditions (Figure 7). Calibration results are also in agreement with qualitative historical data and with previous modeling efforts.

5. THE SWI PACKAGE

M. Bakker

The Sea Water Intrusion (SWI) package is intended for modeling regional seawater intrusion with MODFLOW. The package may be used to simulate the three-dimensional evolution of the salinity distribution, taking density effects into account explicitly. The main advantage of the SWI package is that each aquifer can be modeled with a single layer of cells, without requiring vertical discretization of an aquifer. An existing MODFLOW model of a coastal aquifer can be modified to simulate seawater intrusion with the SWI package through the addition of one input file. The SWI package can simulate interface flow, stratified flow, and continuously varying density flow.

5.1 Theory

The basic idea behind the SWI package is that the groundwater in each aquifer is discretized vertically into a number of zones bounded by curved surfaces. A schematic vertical cross-section of an aquifer is shown in Figure 9a; the thick lines represent the surfaces. The elevation of each surface is a unique function of the horizontal coordinates. The SWI package has two options. For the *stratified flow* option, water has a constant density between surfaces and the surfaces represent interfaces; the density is discontinuous across a surface (Figure 9b). For the *variable density flow* option, the surfaces represent iso-surfaces of the density; the density varies linearly in the vertical direction between surfaces and is continuous across a surface (Figure 9c).

Four main approximations are made:

- The Dupuit approximation is adopted and is interpreted to mean that the resistance to flow in the vertical direction is neglected. The Dupuit approximation is accurate for many practical problems of interface flow, even when the slope of the interface is relatively steep (up to 45°), and for variable density flow [Strack and Bakker, 1995]. The vertical pressure distribution is hydrostatic in each aquifer, but this does not mean that there is no vertical flow; the vertical component of flow is computed from three-dimensional continuity of flow.
- The mass balance equation is replaced by the continuity of flow equation in the computation of the flow field (the Boussinesq-Oberbeck approximation); density effects are taken into account through Darcy's law.
- Effects of dispersion and diffusion are not taken into account.
- Inversion is not allowed. Inversion means that saltier (heavier) water is present above fresher (lighter) water, often resulting in the vertical growth of fingers. The SWI package is intended for the modeling of regional seawater intrusion, which is generally on a scale well beyond the size of the fingers.

Dependent variables in the formulation are the freshwater head at the top of each aquifer and the elevations of the surfaces in each aquifer, and the vertically integrated fluxes. Application of continuity of flow in each aquifer results in a system of differential equations for the freshwater head that is identical in form to the differential equations for single-density flow, but with an additional pseudo-source term, representing the density effects, on the right-hand side (RHS). Hence, MODFLOW can be used to compute the distribution of the freshwater head by addition of this pseudo-source term to the RHS. The differential equations that govern the movement of the surfaces have the same form as the equations for the head, but with different values for the transmissivities and pseudo-source term. Since the form is the same, the solution engines of MODFLOW can again be applied to solve the system for every timestep. A simple tip/toe tracking algorithm is applied to keep track of the horizontal positions of the surfaces. Details of the theory implemented in the SWI package may be found in Bakker [2003].

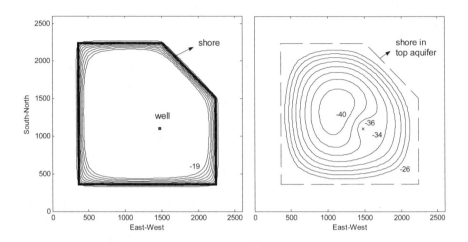

Figure 10: Contours of interface elevation below an island with a well in the top aquifer after 40 years of pumping: top aquifer (left) and bottom aquifer (right).

5.2   Example Application

The SWI package is implemented in MODFLOW2000. Only one additional input file is needed to simulate seawater intrusion. The input file consists of the elevations of the surfaces in each aquifer, the density between the surfaces, whether flow should be treated as stratified or variable density, and some tip/toe tracking parameters. MODFLOW/SWI may then be used to compute the positions of the surfaces at the requested times. Details of application of the SWI package may be found in the manual [Bakker and Schaars, 2003]; an executable, a manual, and the source code are available for free download from the author's web page.[2]

One of the major benefits of the SWI package is that it can simulate interface flow, stratified flow, and variable density flow efficiently, even in the same model. Especially when little data is available, it is useful to determine the steady-state position of the interface. This position may already be sufficient to solve the posed problem, or may be used as a starting point for additional transient simulations. When a significant brackish zone is present, the interface may be replaced by one or more brackish zones, either of constant density or variable density. One aquifer may have an interface, while another may have a brackish zone, as will be demonstrated below.

---

[2] http://www.engr.uga.edu/~mbakker/swi.html

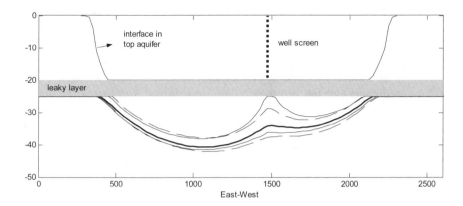

Figure 11: Upconing of brackish zone along east-west cross-section through well: 20 years (dashed), 40 years (solid), interface after 40 years (bold).

Consider seawater intrusion below the hypothetical five-sided island shown in Figure 10. The top aquifer extends from 0 to –20 m, and has a transmissivity of 100 m^2/d; the bottom aquifer extends from –25 to –55 m, and has a transmissivity of 150 m^2/d. The leakance (V_{cont}) of the leaky layer is 0.002 d^{-1}. The island is surrounded by the ocean, with a fixed level of 0 m; the vertical leakance of the bottom of the ocean, representing the vertical resistance to outflow into the ocean, is 0.1 d^{-1}. Recharge on the island is 0.5 mm/d and is specified with the RCH package. The effective porosity of both aquifers is 0.2. The freshwater heads are computed assuming consecutive steady-state conditions, as the heads will react much quicker than the position of the interface; heads can be treated as transient as well, but modeling them as consecutive steady-states has little influence on the results and allows for the specification of much larger timesteps.

The island is discretized into cells of 25 by 25 meters; the grid is extended at least 350 m into the ocean in all directions. The ocean cannot be modeled with GHB cells, as all sinks and sources in the SWI package are treated as consisting of freshwater. The ocean is modeled with an additional layer on top of the model, consisting of inactive cells wherever the island sticks out of the ocean, and fixed head cells elsewhere. Surfaces or interfaces will be specified at the top of the additional layer, such that all water in the additional layer is salt.

As a first step in the modeling process, flow is treated as interface flow. The saltwater has a density of 1,025 kg/m^3. The maximum slope of the interface is specified as 0.03, and the other two tip/toe tracking parameters are specified according to the guidelines in the SWI manual. The steady-state position of the interface is approached after 80 steps of 250 days, starting from a rough first guess. The freshwater zone in the bottom aquifer is over 18 m thick in the middle of the island. The steady-state position is used as a starting point for further modeling. A well is started in the top aquifer and has a discharge of 200 m^3/d (about 10% of total recharge on the island). Contours of the elevation of the interface after 40 timesteps of 1 year are shown in Figure 10. The well has little effect on the position of the interface in the top aquifer, but there is an upconing of 8 m below the well in the bottom aquifer.

As it is crucial for the saltwater to remain in the bottom aquifer and not reach the leaky layer below the well, modeling is continued by replacing the interface in the lower aquifer with a brackish zone, initially extending 5 m above the steady-state position of the interface. The brackish water has a constant density of 1012.5 kg/m^3. The position of the brackish zone along an east-west cross-section through the well is shown after 20 years (dashed) and 40 years (solid) of pumping in Figure 11; results of the interface simulation after 40 years of pumping are also shown in the figure (thick line). It is concluded that the top of a 5 m thick brackish zone will reach the bottom of the leaky layer after 40 years.

6. SUMMARY

This chapter presents four MODFLOW-based codes for simulation of variable-density groundwater flow. An example application was presented for each code to demonstrate the simulation capabilities. Additional information for each code can be found on the accompanying CD of this book.

REFERENCES

Bakker, M., "A Dupuit formulation for modeling seawater intrusion in regional aquifer systems," *Water Resources Research*, in print, 2003.

Bakker, M. and Schaars, F., "The Sea Water Intrusion (SWI) package manual, version 0.2," http://www.engr.uga.edu/~mbakker/swi.html, 2003.

Barry, D.A., Prommer, H., Miller, C.T., Engesgaard, P., and Zheng, C., "Modelling the fate of oxidisable organic contaminants in groundwater," *Adv. Water Resources*, **25**, 899–937, 2002.

Christensen, F.D., Basberg, L., and Engesgaard, P., "Modeling transport and biogeochemical processes in dense landfill leachate plumes," In: Computational Methods in Water Resources, Proceedings of the XIVth International Conference, Delft, Netherlands, June 23–28, 2002.

Christensen, F.D., Engesgaard, P., and Kipp, K.L., "A reactive transport investigation of seawater intrusion experiment in a shallow aquifer, Skansehage, Denmark," Proceedings of the First International Conference on Saltwater Intrusion and Coastal Aquifers, Essaouira, Morocco, April 23–25, 2001.

Clement, T.P., "A modular computer code for simulating reactive multispecies transport in 3-dimensional groundwater systems," Technical report PNNL-SA-11720, Pacific Northwest National Laboratory, Richland, WA, 1997.

Guo, W. and Bennett, G.D., "SEAWAT version 1.1: A computer program for simulations of groundwater flow of variable density," Missimer International, Inc., Fort Myers, FL, 1998.

Guo, W. and Langevin, C.D., "User's guide to SEAWAT: A computer program for simulation of three-dimensional variable-density ground-water flow," U.S. Geological Survey Open-File Report 01-434, 79 p., 2002.

Harbaugh, A.W. and McDonald, M.G., "User's documentation for MODFLOW-96, an update to the U.S. Geological Survey modular finite-difference ground-water flow model," U.S. Geological Survey Open-File Report 96-0485, 56 p., 1996.

Harbaugh, A.W., Banta, E.R., Hill, M.C., and McDonald, M.G., "MODFLOW-2000, the U.S. Geological Survey modular ground-water model—user guide to modularization concepts and the ground-water flow process," U.S. Geological Survey Open-File Report 00-92, 121 p., 2000.

Hill, M.C., Poeter, E., Zheng, C., and Doherty, J., "MODFLOW2001 and other modeling odysseys," *Ground Water*, **41**, 113, 2003.

Hsieh, P.A. and Winston, R.B., "User's guide to Model Viewer, a program for three-dimensional visualization of ground-water model results," U.S. Geological Survey Open-File Report 02-106, 18 p., 2002.

HydroGeoLogic, Inc., "MODHMS—MODFLOW-based Hydrologic Modeling System: Documentation and User's Guide," 2002.

HydroGeoLogic, Inc., "Three-dimensional density-dependent flow and transport modeling of saltwater intrusion in the Southern Water Use Caution Area," Prepared for the Southwest Florida Water Management District, June 2002.

ICW, "Hydrology and water quality of the central part of the western Netherlands," (in Dutch), ICW Regional Studies 9, Institute for Land and Water Management Research, Wageningen: 101 pp., 1976.

IPCC, Intergovernmental Panel on Climate Change. Climate "Change 2001: The Scientific Basis," http://www.ipcc.ch/, 2001.

Konikow, L.F. and Bredehoeft, J.D., "Computer model of two-dimensional solute transport and dispersion in ground-water," U.S. Geological Survey Techniques of Water-Resources Investigation Book 7, Chapter C2, 1978.

Konikow, L.F., Goode, D.J., and Hornberger, G.Z., "A three-dimensional method-of-characteristics solute-transport model (MOC3D)," US Geological Survey Water-Resources Investigations Report 96-4267, 87 p., 1996.

Langevin, C.D., Swain, E.D., and Wolfert, M., "Numerical simulation of integrated surface-water/ground-water flow and solute transport in the southern Everglades in Florida," Presented at the Second Federal Interagency Hydrologic Modeling Conference, Las Vegas, NV, July 28–August 1, 2002.

Langevin, C.D., Swain, E.D., and Wolfert, M., "Flows, stages, and salinities: how accurate is the SICS integrated surface-water/ground-water flow and solute transport model?" In: Florida Bay Program & Abstracts, Joint Conference on the Science and Restoration of the Greater Everglades and Florida Bay Ecosystem, Palm Harbor, FL, April 13–18, 2003.

Lebbe, L.C., "Mathematical model of the evolution of the fresh-water lens under the dunes and beach with semi-diurnal tides," In: Proceedings of the 8th Salt Water Intrusion Meeting, Bari, Italy, May 1983, Geologia Applicata e Idrogeologia, Vol. XVIII, Parte II: p. 211–226, 1983.

Leendertse, J.J., "Aspects of SIMSYS2D, a system for two-dimensional flow computation," Santa Monica, CA, RAND Corp., Report No. R-3572-USGS, 80 p., 1987.

Maas, C. and Emke, M.J., "Solving varying density groundwater problems with a single density program," In: Proceedings of the 10th Salt Water Intrusion Meeting, Ghent, 143–154, 1988.

Mayer, K.U., Frind, E.O., and Blowes, D.W., "Multicomponent reactive transport modeling in variably saturated porous media using a generalized formulation for kinetically controlled reactions," *Water Resources Research*, **38**(9), art. no. 1174, 2002.

McDonald, M.G. and Harbaugh, A.W., "A modular three-dimensional finite difference groundwater flow model," U.S. Geological Survey Techniques of Water Resources Investigations Report, Book 6, Chapter 1, 1988.

Olsthoorn, T.N., "Variable density modelling with MODFLOW," In: Proceedings of the 14th Salt Water Intrusion Meeting, MALMO, 51–58, 1996.

Oude Essink, G.H.P., "MOC3D adapted to simulate 3D density-dependent groundwater flow," In: Proceedings of the MODFLOW 98 Conference, Golden, CO, 291–303, 1998.

Oude Essink, G.H.P., "Impact of sea level rise in the Netherlands," Chap. 14, In: *Seawater Intrusion in Coastal Aquifers—Concepts, Methods and Practices*, (eds.) J. Bear, A.H.-D. Cheng, S. Sorek, D. Ouazar, and I. Herrera, 507–530, 1999.

Oude Essink, G.H.P., "Salt Water Intrusion in a Three-dimensional Groundwater System in The Netherlands: a Numerical Study," *Transport in Porous Media*, **43** (1), 137–158, 2001.

Oude Essink, G.H.P. and Schaars, F., "Impact of climate change on the groundwater system of the water board of Rijnland, The Netherlands," In: Proceedings of the 17th Salt Water Intrusion Meeting, Delft, The Netherlands, May 2002, 379–392, 2003.

Parkhurst, D.L. and Appelo, C.A.J., "User's guide to PHREEQC—A computer program for speciation, reaction-path, 1D-transport, and inverse geochemical calculations," U.S. Geological Survey Water-Resources Investigations Report 99-4259, 1999.

Parkhurst, D.L., Engesgaard, P., and Kipp, K.L., "Coupling the geochemical model PHREEQC with a 3D multi-component solute transport model," In: Fifth Annual V. M. Goldschmidt Conference, Penn State University, University Park, PA, May 24–26, 1995.

Pollock, D.W., "User's Guide for MODPATH/MODPATH-PLOT, Version 3: A particle tracking post-processing package for MODFLOW, the U.S. Geological Survey finite-difference ground-water flow model," U.S. Geological Survey Open-File Report 94-464, 1994.

Prommer, H., Barry, D.A., Chiang, W.H., and Zheng, C., "PHT3D—A MODFLOW/MT3DMS based reactive multi-component transport model," *Ground Water,* **42**(2), 247–257, 2003.

Steefel, C.I., "GIMRT, version 1.2: Software for modeling multicomponent, multidimensional reactive transport. User's Guide," Report UCRL-MA-143182, Lawrence Livermore National Laboratory, Livermore, CA, 2001.

Strack, O.D.L. and Bakker, M., "A validation of a Dupuit-Forchheimer formulation for flow with variable density," *Water Resources Research*, **31**(12), 3019–3024, 1995.

Swain, E.D., Langevin, C.D., and Wolfert, M., "Developing a computational technique for modeling flow and transport in a density-dependent coastal wetland/aquifer system," In: Florida Bay Program & Abstracts, Joint Conference on the Science and Restoration of the Greater Everglades and Florida Bay Ecosystem, Palm Harbor, FL, April 13–18, 2003.

Weiss, E., "A model for the simulation of flow of variable-density ground water in three dimensions under steady-state conditions," U.S. Geological Survey Open-File Report 82-352, 59 p. 1982.

Yeh, G.-T., Salvage, K.M., Gwo, J.P., Zachara, J.M., and Szecsody, J.E., "HydroBioGeoChem: A Coupled Model of Hydrologic Transport and Mixed Biogeochemical Kinetic/Equilibrium Reactions in Saturated-Unsaturated Media," Report ORNL/TM-13668. Oak Ridge National Laboratory, Oak Ridge, TN, 1998.

Yeh, G.-T. and V.S. Tripathi, "A critical evaluation of recent developments of hydrogeochemical transport models of reactive multi-chemical components," *Water Resources Research,* **25**(1), 93–108, 1989.

Zhang, H., Schwartz, F.W., Wood, W.W., Garabedian, S.P., and LeBlanc, D.R. "Simulation of variable-density flow and transport of reactive and non-reactive solutes during a tracer test at Cape Cod, Massachusetts," *Water Resources Research*, **34**(1), 67–82, 1998.

Zheng, C. and Wang, P.P., "MT3DMS, A modular three-dimensional multi-species transport model for simulation of advection, dispersion and chemical reactions of contaminants in groundwater systems; documentation and user's guide," Contract Report SERDP-99-1, U.S. Army Engineer Research and Development Center, Vicksburg, MS, 1999.

CHAPTER 4

Modeling Three-Dimensional Density Dependent Groundwater Flow at the Island of Texel, The Netherlands

G.H.P. Oude Essink

1. INTRODUCTION

Texel is the biggest Dutch Wadden island in the North Sea. It is often called Holland in a nutshell (Figure 1a). The population of the island is about 13,000, whereas in summertime, the number of people can be as high as 60,000. A sand-dune area is present at the western side of the island, with phreatic water levels up to 4 m above mean sea level. At the eastern side, four low-lying polders[1] with controlled water levels are present (Figures 1b and 2a). The lowest phreatic water levels can be measured in the so-called Prins Hendrik polder (reclaimed as tidal area in 1847), with levels as low as –2.0 m N.A.P.[2] In addition, a dune area called De Hooge Berg, which is situated in the southern part of the island in the polder area Dijkmanshuizen, has a phreatic water level of +4.75 m N.A.P. The De Slufter nature reserve in the northwestern part of the island is a tidal salt marsh.

The island of Texel faces a number of water management problems. Agriculture has to deal with salinization of the soils. In nature areas there is not enough water available of sufficient high quality. During summer time, the tourist industry requires large amounts of drinking water while sewage water cannot be easily disposed. In addition, climate change and sea level rise will increase the stresses on the whole water system. On the average, the freshwater resources at the island are too limited to structurally solve these above-mentioned problems.

Therefore, the consulting engineering company Witteveen & Bos executed a study, called "Great Geohydrological Research Texel," to analyze

---

[1] A polder is an area that is protected from water outside the area, and that has a controlled water level.

[2] N.A.P. stands for Normaal Amsterdams Peil. It roughly equals Mean Sea Level and is the reference level in The Netherlands.

1-56670-605-X/04/$0.00+$1.50
© 2004 by CRC Press LLC

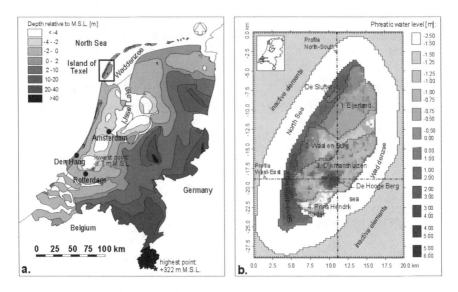

Figure 1: (a) Map of The Netherlands: position of the island of Texel and ground surface of The Netherlands; (b) map of Texel: position of the four polder areas and sand-dune area as well as phreatic water level in the top aquifer at –0.75 m N.A.P. The polder area Eijerland was retrieved from the tidal planes and created during the years 1835–1876. The two profiles refer to Figures 8 and 9.

these water management problems and to gain a comprehensive, coherent knowledge about the whole water system. In addition, technical measures were suggested to control water management in the area. In this article, the interest is only focused on a part of the study, viz. the density-driven groundwater system under changing environmental conditions. The author of this article constructed the density-driven groundwater system with the help of Jeroen Tempelaars and Arco van Vugt.

First, the computer code, which is used to simulate variable density flow in this groundwater system, is summarized. Second, the model of Texel will be designed, based on subsoil parameters, model parameters, and boundary conditions. The numerical results of the autonomous situation and one scenario of sea level rise are discussed in the next section, and finally, conclusions are drawn.

2. CHARACTERISTICS OF THE NUMERICAL MODEL

MOCDENS3D [Oude Essink, 1998] is used to simulate the transient groundwater system as it occurs on the island of Texel. Originally, this code was the three-dimensional computer code MOC3D [Konikow *et al.*, 1996].

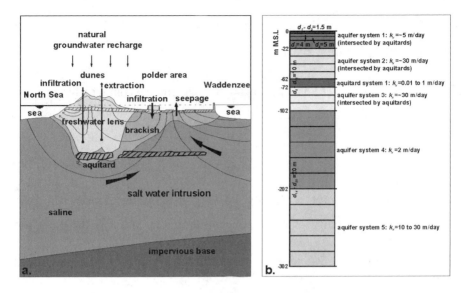

Figure 2: (a) A schematization of the hydrogeological situation at the island of Texel, The Netherlands; (b) the simplified composition of the subsoil into six main subsystems: one aquitard system and five aquifer systems (of which the top three are intersected by aquitards).

2.1   Groundwater Flow Equation

The MODFLOW module solves the density-driven groundwater flow equation [McDonald and Harbaugh, 1988; Harbaugh and McDonald, 1996]. It consists of the continuity equation combined with the equation of motion. Under the given circumstances in the Dutch coastal aquifers, the Oberbeck-Boussinesq approximation is valid as it is suggested that the density variations (due to concentration changes) remain small to moderate in comparison with the reference density ρ throughout the considered hydrogeologic system:

$$\frac{\partial q_x}{\partial x} + \frac{\partial q_y}{\partial y} + \frac{\partial q_z}{\partial z} = S_s \frac{\partial \phi_f}{\partial t} + W \tag{1}$$

$$q_x = -\frac{\kappa_x}{\mu}\frac{\partial p}{\partial x}; \qquad q_y = -\frac{\kappa_y}{\mu}\frac{\partial p}{\partial y}; \qquad q_z = -\frac{\kappa_z}{\mu}\left(\frac{\partial p}{\partial z} + \rho g\right) \tag{2}$$

where q_x, q_y, q_z = Darcian specific discharges in the principal directions $[LT^{-1}]$; S_s = specific storage of the porous material $[L^{-1}]$; W = source function, which describes the mass flux of the fluid into (negative sign) or out of (positive sign) the system $[T^{-1}]$; κ_x, κ_y, κ_z = principal intrinsic permeabilities $[L^2]$; μ = dynamic viscosity of water $[ML^{-1}T^{-1}]$; p = pressure $[ML^{-1}T^{-2}]$; and g = gravitational acceleration $[LT^{-2}]$. A so-called freshwater head ϕ_f $[L]$ is introduced to take into account differences in density in the calculation of the head:

$$\phi_f = \frac{p}{\rho_f g} + z \tag{3}$$

where ρ_f = the reference density $[ML^{-3}]$, usually the density of fresh groundwater at reference chloride concentration C_0, and z is the elevation head $[L]$.

Rewriting the Darcian specific discharge in terms of freshwater head gives:

$$q_x = -\frac{\kappa_x \rho_f g}{\mu}\frac{\partial \phi_f}{\partial x}; \qquad\qquad q_y = -\frac{\kappa_y \rho_f g}{\mu}\frac{\partial \phi_f}{\partial y};$$

$$q_z = -\frac{\kappa_z \rho_f g}{\mu}\left(\frac{\partial \phi_f}{\partial z} + \frac{\rho - \rho_f}{\rho_f}\right) \tag{4}$$

In many cases small viscosity differences can be neglected if density differences are considered in normal hydrogeologic systems [Verruijt, 1980; Bear and Verruijt, 1987].

$$\frac{\kappa_i \rho_f g}{\mu} = k_i \tag{5}$$

$$q_x = -k_x \frac{\partial \phi_f}{\partial x}; \quad q_y = -k_y \frac{\partial \phi_f}{\partial y}; \quad q_z = -k_z\left(\frac{\partial \phi_f}{\partial z} + \frac{\rho - \rho_f}{\rho}\right) \tag{6}$$

The basic water balance used in MODFLOW is given below [McDonald and Harbaugh, 1988]:

$$\sum_i Q_i = S_s \frac{\Delta \phi_f}{\Delta t} \Delta V \tag{7}$$

where Q_i = total flow rate into the element ($L^3 T^{-1}$) and ΔV = volume of the element (L^3). The MODFLOW basic equation for density dependent groundwater flow becomes as follows [Oude Essink, 1998, 2001]:

$$CV_{i,j,k-\frac{1}{2}} \phi_{i,j,k-1}^{t+\Delta t} + CC_{i-\frac{1}{2},j,k} \phi_{i-1,j,k}^{t+\Delta t} + CR_{i,j-\frac{1}{2},k} \phi_{i,j-1,k}^{t+\Delta t}$$

$$- (CV_{i,j,k-\frac{1}{2}} + CC_{i-\frac{1}{2},j,k} + CR_{i,j-\frac{1}{2},k}$$

$$+ CV_{i,j,k+\frac{1}{2}} + CC_{i+\frac{1}{2},j,k} + CR_{i,j+\frac{1}{2},k} - HCOF_{i,j,k}) \phi_{i,j,k}^{t+\Delta t}$$

$$+ CV_{i,j,k+\frac{1}{2}} \phi_{i,j,k+1}^{t+\Delta t} + CC_{i+\frac{1}{2},j,k} \phi_{i+1,j,k}^{t+\Delta t} + CR_{i,j+\frac{1}{2},k} \phi_{i,j+1,k}^{t+\Delta t} = RHS_{i,j,k}$$

$$HCOF_{i,j,k} = P_{i,j,k} - SC1/\Delta t$$

$$RHS_{i,j,k} = -Q_{i,j,k} - SC1_{i,j,k} \phi_{i,j,k}^{t} / \Delta t$$
$$- CV_{i,j,k-\frac{1}{2}} \Psi_{i,j,k-\frac{1}{2}} (d_{i,j,k-1} + d_{i,j,k})/2$$
$$+ CV_{i,j,k+\frac{1}{2}} \Psi_{i,j,k+\frac{1}{2}} (d_{i,j,k} + d_{i,j,k+1})/2$$

$$SC1_{i,j,k} = SS_{i,j,k} \Delta V \tag{8}$$

$$\Psi_{i,j,k-1/2} = \left(\frac{(\rho_{i,j,k-1} + \rho_{i,j,k})/2 - \rho_f}{\rho_f} \right)$$

$$\Psi_{i,j,k+1/2} = \left(\frac{(\rho_{i,j,k} + \rho_{i,j,k+1})/2 - \rho_f}{\rho_f} \right) \tag{9}$$

where $CV_{i,j,k}, CC_{i,j,k}, CR_{i,j,k}$ = the so-called MODFLOW hydraulic conductance between elements in respectively vertical, column, and row directions ($L^2 T^{-1}$) [McDonald and Harbaugh, 1988]; $P_{i,j,k}, Q_{i,j,k}$ = factors that account for the combined flow of all external sources and stresses into an element ($L^2 T^{-1}$); $SS_{i,j,k}$ = specific storage of an element (L^{-1}); $d_{i,j,k}$ = thickness of the model layer k (L), and $\Psi_{i,j,k}$ = buoyancy terms (dimensionless). The two buoyancy terms $\Psi_{i,j,k}$ are subtracted from the so-called right head side term $RHS_{i,j,k}$ to take into account variable density. See Oude Essink [1998, 2001] for a detailed description of the adaptation of MODFLOW to density differences.

2.2 The Advection-Dispersion Equation

The MOC module uses the method of characteristics to solve the advection-dispersion equation, which simulates the solute transport [Konikow and Bredehoeft, 1978; Konikow et al., 1996]. Advective transport of solutes is modeled by means of the method of particle tracking and dispersive transport by means of the finite difference method:

$$R_d \frac{\partial C}{\partial t} = \frac{\partial}{\partial x_i} \left(D_{ij} \frac{\partial C}{\partial x_j} \right) - \frac{\partial}{\partial x_i} \left(CV_i \right) + \frac{(C-C)'W}{n_e} - R_d \lambda C \qquad (10)$$

The used reference solute is chloride that is expected to be conservative. MOCDENS3D takes into account hydrodynamic dispersion.

2.3   The Equation of State

A linear equation of state couples groundwater flow and solute transport:

$$\rho_{i,j,k} = \rho_f [1 + \beta_C (C - C_0)] \qquad (11)$$

where $\rho_{i,j,k}$ is the density of groundwater (ML^{-3}), C is the chloride concentration (ML^{-3}), and β_C is the volumetric concentration expansion gradient ($M^{-1}L^3$). During the numerical simulation, changes in solutes, transported by advection, dispersion, and molecular diffusion, affect the density and thus the groundwater flow. The groundwater flow equation is recalculated regularly to account for changes in density.

2.4   Examples of Three-Dimensional Studies with MOCDENS3D

The computer code MOCDENS3D has recently also been used for three other three-dimensional regional groundwater systems in The Netherlands: (a) the northern part of the province of North-Holland: 65.0 km by 51.25 km by 290 m with ~40,000 active elements [Oude Essink, 2001]; (b) the Wieringermeerpolder at the province of North-Holland: 23.2 km by 27.2 km by 385 m with ~312,000 active elements [Oude Essink, 2003; Water board Uitwaterende Sluizen, 2001]; and (c) the water board of Rijnland in the province of South-Holland: 52.25 km by 60.25 km by 190 m with 1,209,000 active elements [Oude Essink and Schaars, 2003; Water Board of Rijnland, 2003].

3.   MODEL DESIGN

3.1   Geometry, Model Grid, and Temporal Discretization

The following parameters are applied for the numerical computations. The groundwater system consists of a three-dimensional grid of 20.0 km by 29.0 km by 302 m depth. Each element is 250 m by 250 m long. In vertical direction the thickness of the elements varies from 1.5 m at the top layer to 20 m over the deepest 10 layers (Figure 2b). The grid contains 213,440 elements: $n_x = 80$, $n_y = 116$, $n_z = 23$, where n_i denotes the number of elements in the i direction. Due to the rugged coastline of the system and the irregular shape of the impervious hydrogeologic base, only 58.8% of the elements (125,554 out of 213,440) are considered as active elements. Each active element contains eight particles to solve the advection term of the solute transport equation. As such, some one million particles are used initially. The flow timestep Δt to recalculate the groundwater flow equation equals 1 year. The convergence criterion for the groundwater flow equation (freshwater head) is equal to 10^{-5} m. The total simulation time is 500 years.

3.2   Subsoil Parameters

The groundwater system consists of permeable aquifers, intersected by loamy aquitards and aquitards of clayey and peat composite (Figure 2b). The system can be divided into six main subsystems. The top subsystem (from 0 m to –22 m N.A.P.) and the second subsystem (from –22 m to –62 m N.A.P.) have hydraulic conductivities k_x of approximately 5 m/d and 30 m/d, respectively. The third subsystem is an aquitard of 10 m thickness and has hydraulic conductivities k_x that varies from 0.01 to 1 m/d. The fourth subsystem (from –72 m to –102 m N.A.P.) and fifth subsystem (from –102 m to –202 m N.A.P.) have hydraulic conductivities k_x of some 30 m/d and only 2 m/d, respectively. The lowest subsystem, number six, has a hydraulic conductivity k_x of approximately 10 m/d to 30 m/d. Note that the first, second, and fourth subsystems are intersected by aquitards.

The following subsoil parameters are assumed: the anisotropy ratio k_z/k_x equals 0.4 for all layers. The effective porosity n_e is 0.35. The longitudinal dispersivity α_L is set equal to 2 m, while the ratio of transversal to longitudinal dispersivity is 0.1. For a conservative solute as chloride, the molecular diffusion for porous media is taken equal to 10^{-9} m^2/s. Note that no numerical "Peclet" problems occurred during the simulations [Oude Essink and Boekelman, 1996]. On the applied time scale, the specific storativity S_s (L^{-1}) can be set to zero.

The bottom of the system as well as the vertical seaside borders is considered to be no-flux boundaries. At the top of the system, the mean sea level is –0.10 m N.A.P. and is constant in time in case of no sea level rise.[3]

---

[3] Note that in reality, the mean sea level in the eastern direction toward the Waddenzee is probably somewhat higher over a few hundreds of meters. The reason is that at low tide, the piezometric head in the phreatic aquifer of this tidal foreland outside the dike cannot follow the relatively rapid tidal surface water fluctuations (Lebbe, pers. comm., 2000). It will be retarded, which results in a higher low tide level of the sea, and thus in a higher mean sea level.

A number of low-lying areas are present in the system with a total area of approximately 124 km^2. The phreatic water level in the polder areas differs significantly, varying from –2.05 m to +4.75 m N.A.P. at the hill De Hooge Berg (Figure 1b), and is kept constant in time. Small fluctuations in the phreatic water level are neglected. The constant natural groundwater recharge equals 1 mm/d in the sand-dune area.

The volumetric concentration expansion gradient β_C is 1.34×10^{-6} l/mg Cl$^-$. Saline groundwater in the lower layers does not exceed 18,000 mg Cl$^-$/l, as seawater that intruded the groundwater system has been mixed with water from the river Rhine. The corresponding density of that saline groundwater equals 1024.1 kg/m^3.

3.3   Determination of the Initial Density Distribution

By 1990 AD, the hydrogeologic system contains saline, brackish as well as fresh groundwater. On the average, the salinity increases with depth, whereas freshwater lenses exist at the sand-dune areas at the western side of the island, up to some –50 m N.A.P. A freshwater lens of some 50 m thickness has evolved at the sandy hill De Hooge Berg.

Head as well as density differences affect groundwater flow in this system. Density-driven groundwater flow simulated with a numerical model is very sensitive to the accuracy of the initial density distribution. As such, the initial chloride concentration, which is linearly related to the initial density by Eq. (10), must be accurately inserted in each active element.

In this particular situation,[4] the present density distribution cannot be deduced by simply simulating the saline groundwater system for many hundreds of years with all actual load and concentration boundary conditions, and waiting until the composition of solutes ceases to change. The reason is that the present distribution of fresh, brackish, and saline groundwater is still not in equilibrium. Several processes initiated in the past can still be sensed and make the situation dynamic. For instance, during the past centuries, the position of the island of Texel itself was not fixed [Province of North-Holland, 2000]. It has slowly been moved, mainly from the west to the east [Oost, 1995]. As a consequence, freshwater lenses in the sand-dune areas could not follow the moving upper boundary conditions of natural groundwater recharge. Moreover, other human activities such as polders were created, some even from the 17th century on. Groundwater extractions confirm the dynamic character of the island.

Therefore, from a practical point of view and based on the fact that the system is still dynamic, chloride (and thus density) measurements at the

---

[4] As a matter of fact, the same circumstances are present in most other coastal areas in The Netherlands.

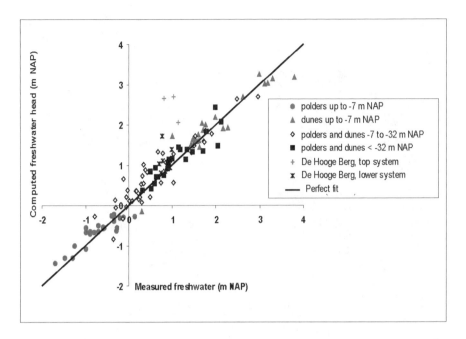

Figure 3: Calibration of the freshwater head: computed versus "measured" freshwater heads.

year 1990 AD are chosen as the initial situation. Though this initial chloride distribution in this Texel case is based on about 100 measurements of chloride, errors can easily occur, mainly because of a lack of enough data. Artificial inversions of fresh and saline groundwater can easily occur in the numerical model, though they do not exist in reality. As a remedy, 10 years are simulated under reference conditions (e.g., constant head at polders and the sea), viz. from 1990 to 2000 AD. These years are necessary to smooth out unwanted, unrealistic density dependent groundwater flow, which was caused by the numerical discretization of the initial density distribution.

4. DISCUSSION

4.1 Calibration of the Model

Calibration of the numerical model was focused on the freshwater heads in the hydrogeologic system, as well as on seepage and salt load values that were measured at five pumping stations in the surface water system [Province of North-Holland, 2000]. Freshwater head calibration was executed by comparing 111 measured and simulated (freshwater) heads,

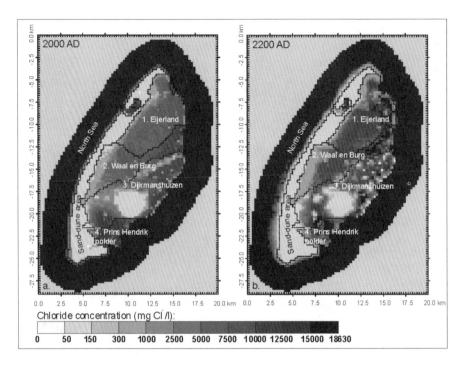

Figure 4: Chloride concentration in the top layer at −0.75 m N.A.P. for the
years 2000 and 2200 AD. No sea level rise is simulated.

which were corrected for density differences. Figure 3 shows the head
calibration. The module PEST of PMWIN (version 5.0) was used to
minimize the difference between measured and simulated (freshwater) heads.
A sensitivity analysis has been executed on the following, in this system,
most important subsoil parameters: drainage resistance; streambed resistance
of the main water channels; vertical resistance of the Holocene aquitard in
the polder area; horizontal hydraulic conductivity of the phreatic aquifer in
the sand-dune area; and the vertical hydraulic conductivities of the aquitards
in aquifer systems two and three (see Figure 2b).

For all observation wells, the mean error was +0.07 m, the mean
absolute error 0.24 m, and the root mean square error 0.36 m. Systematic
errors were not assumed. Seasonal variations in natural recharge obstruct
easy calibration of the density dependent groundwater flow model with
seepage and salt load values.

Overall, more accurate model parameters, e.g., the increase of the
initial number of particles per element, a smaller timestep to recalculate the
velocity field, and a smaller convergence criterion for the groundwater flow
equation, did not significantly improve the numerical simulation of the
salinization process in the hydrogeologic system.

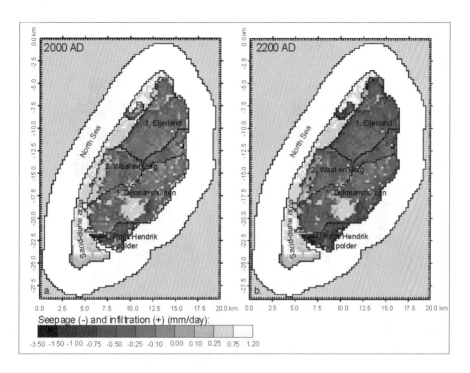

Figure 5: Seepage through the top layer at −0.75 m N.A.P. for the years 2000 and 2200 AD. Sea level rise is 0.75 meter per century.

4.2   Autonomous Saltwater Intrusion during the Period 2000–2200 AD

In the year 2000 AD, the chloride concentration is already high in the four polder areas (Figure 4). At the hill De Hooge Berg, fresh groundwater occurs up to some −45 m M.S.L. Freshwater from the sand-dunes flows toward the sea as well as toward the low-lying polder areas. In these low-lying areas, seepage is quite high (up to some 2.1 mm/day at the western side of the Prins Hendrik polder, see Figure 5a). In addition, the salt load is high too, with values up to some 95,000 kg/ha/year in the same polder area (Figure 6a).

The future autonomous salinization of the groundwater system of the island of Texel is visualized in Figure 4. It shows the change in chloride concentration in the top layer in the years 2000 AD and 2200 AD. The level of sea is kept constant during these 200 years. The salinity in the top layer increases, especially in the areas close to the coastline. The polders, which were created at least 125 years ago, cause the salinity increase. The time lag

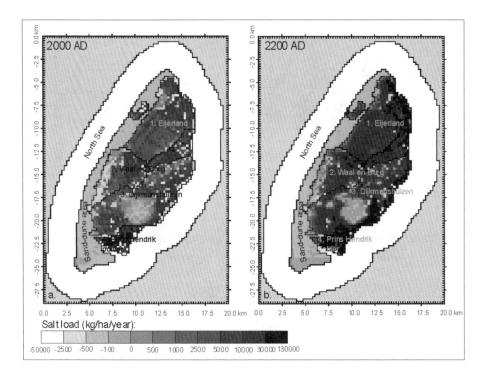

Figure 6: Salt load (in kg/ha/year) in the top layer at −0.75 m N.A.P. for the
years 2000 and 2200 AD. Sea level rise is 0.75 meter per century.

of the salinization process is considerable, at least many tens of years. The
animation of the concentration evolution is provided on the accompanying
CD.

4.3   Effect of Sea Level Rise on Saltwater Intrusion during the Next 500 Years

According to the Intergovernmental Panel of Climate Change
(IPCC) Second Assessment Report [Warrick *et al.*, 1996], a sea level rise of
0.49 m is to be expected for the year 2100, with an uncertainty range from
0.20 to 0.86 m. This rate is 2 to 5 times the rate experienced over the last
century. One scenario of relative sea level variation is considered for the next
500 years: a relative sea level rise of 0.75 m per century. This figure includes
land subsidence caused by groundwater recovery, the compaction and
shrinkage of clay, and especially the oxidation of peat.

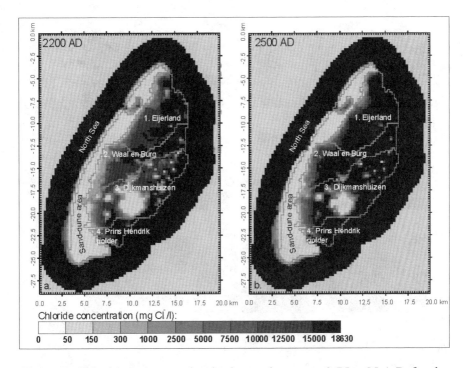

Figure 7: Chloride concentration in the top layer at −0.75 m N.A.P. for the years 2200 and 2500 AD. Sea level rise is 0.75 m per century.

Figure 7 shows the change in chloride concentration in the top layer at −0.75 m N.A.P. for the sea level rise scenario at two moments in time: 200 years (2200 AD) and 500 years (2500 AD) after 2000 AD. During the next centuries, the salinity in the groundwater system will increase very seriously when the sea level rises by 0.75 m/c.

The same process can obviously be detected in a cross-section; see Figure 8 (west-east direction) and Figure 9 (north-south direction). The exact positions of the profiles in these figures are given in Figure 1b. The effect of a sea level rise relative to no sea level rise on the chloride concentration in the top layer can be deduced by comparing Figures 4b and 7a: the salinity increases more rapidly in the low-lying polder areas. The freshwater lenses at the sand-dune area as well as at the hill De Hooge Berg remain, though these lenses become less deep.

Seepage in the polders (Figure 5) as well as the salt load (Figure 6) at −0.75 m N.A.P. increase as a function of time. Two processes cause the increase of salt load: the increase of seepage as well as the increase in salinity of the top hydrogeologic system. The polder areas attract seawater, with a high content of chloride, as the phreatic water level in these areas is low relative to the level of the sea.

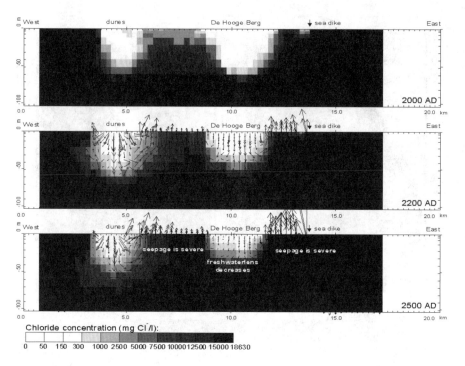

Figure 8: Chloride concentration in a cross-section in western-eastern direction (row 76) for the years 2000, 2200, and 2500 AD, over the hill De Hooge Berg. Only the top system up to −102 m N.A.P. is shown. The relative sea level rise is 0.75 m per century. The arrows correspond with the displacement of groundwater during a time step of 20 years.

In Figure 10, the seepage in the four different polder areas is given as a function of time. As can be seen, seepage quantities increase, which will probably have its effect on the capacity of the pumping stations in the polder areas. Their capacity should be increased because, e.g., in 200 years, the seepage quantity is about doubled in all four areas.

The salt load as a function of time demonstrates that the effect of sea level rise is substantial in all four low-lying polder areas of the island of Texel (Figure 11). The increase in salt load will be enormous due to the sea level rise of 0.75 m per century. This will definitely affect environmental aspects. A doubling of the salt load is probably already reached within (only) one century in the polder areas Eijerland and Dijksmanhuizen.

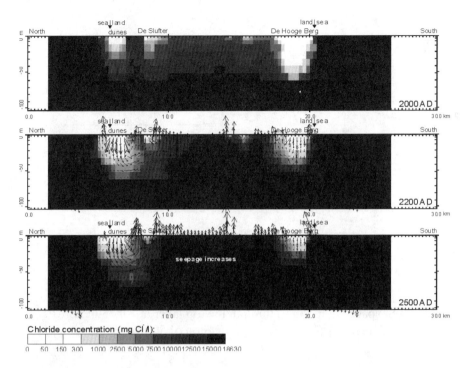

Figure 9: Chloride concentration in a cross-section in northern-southern direction (column 45) for the years 2000, 2200, and 2500 AD. Only the top system up to −102 m N.A.P. is shown. The relative sea level rise is 0.75 m per century. The arrows correspond with the displacement of groundwater during a time step of 20 years.

5. CONCLUSIONS

The "Great Geohydrological Research Texel" was initiated to investigate the effect of environmental and anthropogenic stresses on the groundwater system at the island of Texel. Differences in present water level between the sea and low-lying polders of the island of Texel suggest a large inflow of seawater toward the land. A numerical model was constructed to quantify this phenomenon and to assess the effect of future physical stresses such as sea level rise and land subsidence on the groundwater system. The computer code MOCDENS3D was used to simulate density dependent groundwater flow at the island of Texel in three dimensions with a surface of 130 km^2 by 300 m thickness. The reliability of the numerical model highly depends on the quality of especially the initial density distribution. Numerical computations show that saltwater intrusion is severe because the

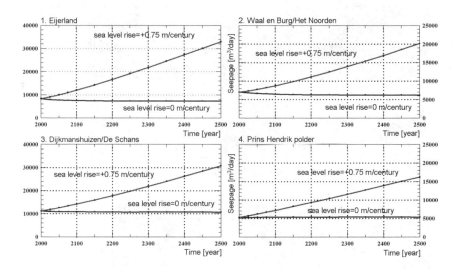

Figure 10: Seepage (in m3/day) through the top layer at −0.75 m N.A.P. of the four polder areas as a function of 500 years.

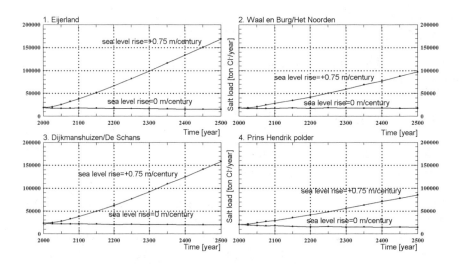

Figure 11: Salt load (in ton Cl⁻/year) through the top layer at −0.75 m N.A.P. of the four polder areas as a function of 500 years.

polder areas with low phreatic water levels are situated very close to the sea. When the sea level rises relatively 0.75 m per century, the increase in salinity is enormous. A doubling of the present seepage quantities can be established within two centuries in all four polder areas. Moreover, the salt load will probably be doubled in two polder areas within only one century. This will definitely affect environmental, as well socio-economic aspects of the island of Texel.

Acknowledgments

The author wishes to thank Jeroen Tempelaars and Arco van Vugt of the consulting engineering company Witteveen & Bos, The Netherlands, for the preparation of the input files (especially subsoil parameters) for the numerical model, as well as executing the sensitivity analysis of subsoil parameters.

REFERENCES

Bear, J. and Verruijt, A., *Modeling Groundwater Flow and Pollution*, D. Reidel Publishing Company, Dordrecht, The Netherlands, 414 p., 1987.

Harbaugh, A.W. and McDonald, M.G., User's documentation for the U.S.G.S. modular finite-difference ground-water flow model, U.S.G.S. Open-File Report 96-485, 56 p., 1996.

Water board Uitwaterende Sluizen, "Geohydrological Research Wieringerrandmeer", by the consulting engineering company Grontmij Noord-Holland, on behalf of the Water board Uitwaterende Sluizen by order of the steering committee "Water Bindt", 48 p., 2001.

Konikow, L.F. and Bredehoeft, J.D., Computer model of two-dimensional solute transport and dispersion in ground water; U.S.G.S. Tech. of Water-Resources Investigations, Book 7, Chapter C2, 90 p., 1978.

Konikow, L.F., Goode, D.J., and Hornberger, G.Z., A three-dimensional method-of-characteristics solute-transport model (MOC3D); U.S.G.S. Water-Resources Investigations Report 96-4267, 87 p., 1996.

McDonald, M.G. and Harbaugh, A.W., A modular three-dimensional finite-difference ground-water flow model; U.S.G.S. Techniques of Water-Resources Investigations, Book 6, Chapter A1, 586 p., 1988.

Oost, A.P., "Dynamics and sedimentary development of the Dutch Wadden Sea with emphasis on the Frisian Inlet", Ph.D. dissertation, Utrecht University, 445 p., 1995.

Oude Essink, G.H.P., "MOC3D adapted to simulate 3D density-dependent groundwater flow," In: Proc. MODFLOW'98 Conf., Golden, CO, 291–303, 1998.

Oude Essink, G.H.P., "Density dependent groundwater flow at the island of Texel, The Netherlands" In: Proc. 16th Salt Water Intrusion Meeting, Miedzyzdroje-Wolin Island, Poland, June 2000, 47–54, 2001.

Oude Essink, G.H.P., "Salt Water Intrusion in a Three-dimensional Groundwater System in The Netherlands: a Numerical Study," *Transport in Porous Media*, **43** (1), 137–158, 2001.

Oude Essink, G.H.P., "Salinization of the Wieringermeerpolder, The Netherlands" In: Proc. 17th Salt Water Intrusion Meeting, Delft, The Netherlands, 399–411, 2003.

Oude Essink, G.H.P. and Schaars, F., "Impact of climate change on the groundwater system of the water board of Rijnland, The Netherlands" In: Proc. 17th Salt Water Intrusion Meeting, Delft, The Netherlands, 379–392, 2003.

Oude Essink, G.H.P. and Boekelman, R.H., "Problems with large-scale modeling of salt water intrusion in 3D," In: Proc. 14th Salt Water Intrusion Meeting, Malmö, Sweden, June 1996, 16–31, 1996.

Province of North-Holland, "Great Geohydrological Research Texel", by the consulting engineering company Witteveen & Bos, on behalf of the Province of North-Holland, the Water board Hollands Kroon, the city Texel and the Water board Uitwaterende Sluizen, 73 p., 2000.

Verruijt, A., "The rotation of a vertical interface in a porous medium," *Water Resour. Res.*, **16** (1), 239–240, 1980.

Warrick, R.A., Oerlemans, J., Woodworth, P., Meier, M.F., and Le Provost, C., "Changes in sea level," In: Climate Change 1995: The Science of Climate, eds. J.T. Houghton, L.G. Meira Filho, and B.A. Callander, Contribution of Working Group I to the Second Assessment Report of the Intergovernmental Panel of Climate Change, 359–405, Cambridge Univ. Press, Cambridge, 1996.

Water Board of Rijnland, "The salt of the earth", by KIWA research and consultancy, on behalf of the Water Board of Rijnland, 2003.

CHAPTER 5

Leaky Coastal Margins: Examples of Enhanced Coastal Groundwater and Surface-Water Exchange from Tampa Bay and Crescent Beach Submarine Spring, Florida, USA

P.W. Swarzenski, J.L. Kindinger

1. INTRODUCTION

As populations and industry migrate toward sought-after coastal zone real estate, increased pressure on these fragile margins demands a realistic and comprehensive understanding of the underlying hydrogeological framework. One of the most threatened resources along these coastal corridors is groundwater, and coastal management agencies have developed complex strategies to protect these resources from overexploitation and contamination. Obvious consequences of coastal groundwater mismanagement may include accelerated saltwater intrusion into supply aquifers, inadequate groundwater supply versus demand, and infiltration of organic and inorganic contaminants into aquifers. Two examples of proactive management strategies in direct response to threatened coastal groundwater resources include the construction and maintenance of injection barrier wells [Johnson and Whitaker, this volume], and the construction of large-scale desalinization plants, such as in Tampa Bay, Florida [Beebe, 2000].

Leaky coastal margins, where exchange processes at the land–sea boundary are naturally enhanced, can include the following environments: *i*) carbonate platforms, *ii*) modern and paleo river channels, *iii*) geothermal aquifers, *iv*) shorelines that are mountainous or have large tidal amplitudes or potentiometric gradients, and *v*) lagoons, where evaporation can force density-driven exchange (Figure 1). In these coastal environments, facilitated fluid–solute exchange can play an important role not only for coastal groundwater/surface water management (i.e., water budgets), but also in the delivery of recently introduced contaminants to coastal bottom waters. This submarine input for nutrients and other waterborne constituents may contribute to coastal eutrophication and other deleterious estuarine impacts.

1-56670-605-X/04/$0.00+$1.50
© 2004 by CRC Press LLC

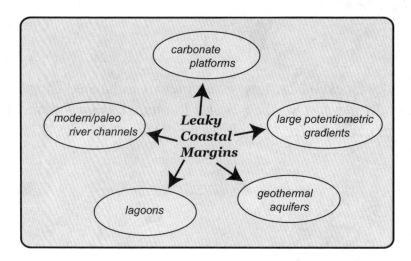

Figure 1: A cartoon depicting some leaky coastal margins.

Such effects can exhibit a full range in scale from being highly localized, for example around a point discharge, to an eventual ecosystem wide shift.

This chapter will discuss some hydrogeologic characteristics unique to leaky coastal margins, and will then illustrate these features by examining two examples from Florida: Tampa Bay and Crescent Beach submarine spring. At each of these sites coastal groundwater resource issues form a critical component in overall ecosystem health, which demands a vigorous interdisciplinary science curriculum.

1.1   Leaky Coastal Margins—Characteristics and Definitions

Thomas [1952] reminded us that the principles of hydrology would be quite simple if the earth's surface could be considered impervious. Components of the water budget would thus be a simple function of precipitation, runoff, and evaporation/transpiration without all the complications of hard to constrain rock–water interactions. We know, however, that water does indeed infiltrate the earth's surface layer. Once a water parcel has been absorbed into subsurface strata, it can accumulate, flow through, be involved in chemical transformation reactions, and eventually discharged. The ability of these strata to hold and transport groundwater depends on the nature of the bedrock and sediments as well as any post-depositional alteration such as faults and dissolution features. The underlying hydrogeologic framework of leaky coastal margins exhibits such subsurface features that directly enhance groundwater transport across a land–sea boundary. This section describes some of the most prevalent coastal depositional environments where such exchange is facilitated.

1.1.1   Carbonate Platforms

Along land–sea margins, limestone, which consists largely of calcite produced by marine organisms, plays a fundamental role in the delicate balance of geologic and biologic cycles. Limestone is biogeochemically reactive as groundwater slowly percolates through interstitial pores and lattices. Dissolution of carbonate rock is caused principally by reactions with water undersaturated in calcium carbonate or acidic water, and will result in pore space enlargements, conduit formation, or large-scale cavities. Dissolution/collapse features such as sinkholes provide direct hydrologic communication between groundwater and surface water and can greatly facilitate water exchange within leaky coastal margins. Often, this facilitated exchange across the sediment–water interface makes it difficult to geochemically distinguish between groundwater and surface water. Along carbonate land–sea margins, the ubiquity of onshore and offshore springs further emphasizes the geologically enhanced water and solute exchange.

1.1.2   Modern and Paleo River Channels

As rivers flow seaward, fluvial processes such as discharge and turbulence continuously sort particles in both the bed and suspended load. As a consequence, paleo and modern river channels are typically well sorted and consist of coarser grained particles such as sands and silts. When a stream or river extends into its adjacent bed or banks, this exchange is considered to occur in the hyporheic zone, and provides a mechanism for the dynamic mixing of groundwater and surface water. Fluctuations in sea level may play an important role in the historic delivery and trajectory of off-continent riverine materials. Coastal riverbeds are therefore an important potential hydrostratigraphic conduit for enhanced groundwater transport offshore. Modern as well as paleo river channels along the eastern seaboard of the United States offer examples of such enhanced exchange.

1.1.3   Geothermal Aquifers

Most work on marine geothermal vents has focused on dramatic open ocean vent systems that are typically basaltic in origin, such as the Galapagos spreading center [Edmond *et al.*, 1979] or the high temperature submarine springs off Baja, California [Vidal *et al.*, 1978]. In Florida, Kohout and colleagues (cf. [Kohout, 1965]) have postulated a geothermally regulated process whereby cold, deep seawater can migrate into the highly permeable layers of the deep Floridan aquifer. Here this water is heated during upward transport and eventually discharged as warm, saline submarine spring water [Fanning *et al.*, 1981]. Because coastal carbonate platforms are fairly common geologic features and as no intense magmatic heat source is required to drive such submarine discharge, the flux of heated groundwater from limestone deposits is likely to be widespread and large enough to affect localized oceanic budgets.

1.1.4   Large Potentiometric Gradients

For many decades, groundwater hydrologists have studied the dynamic transition zone that separates freshwater from saltwater along coastal margins to better predict saltwater intrusion as a potential groundwater contaminant and to more accurately assess the quantity of fresh coastal groundwater.   A general observation from such studies is that the interface in coastal aquifers tends to dip landward due to the increased density of seawater over freshwater, and that the saltwater tongue often extends inland for considerable distances. Another characteristic inherent in any model of this interface, i.e., Badon-Ghijben-Herzberg, Glover [1959], Edelman [1972], Henry [1964], Mualem and Bear [1974], and Meisler et al. [1984], is the direct dependence of the extent of submarine groundwater discharge on elevated potentiometric heads measured at the coast.   For example, on the northern Atlantic coastal margin, where shoreline potentiometric heads were estimated at 6 m, freshwater was modeled to extend about 60 km offshore [Meisler et al., 1984]. Indeed, further south off the coast of northern Florida, freshened groundwater masses were observed to discharge directly into Atlantic bottom waters [Swarzenski et al., 2001]. It is likely that many of these freshened submarine paleo-groundwater masses formed during the Pleistocene when sea levels were lower than at present. This suggests that trapped paleo-groundwaters beneath continental shelfs and shallow seas could provide a substantial groundwater resource, if these deposits could be tapped before processes of natural seawater infiltration contaminate them.

1.1.5   Lagoons

Lagoons are shore-parallel river-ocean mixing zones that are typically developed by marine wave action as opposed to the more traditional river dominated processes that form a deltaic estuary.   Lagoons are often shallow and poorly drained and as a result, water mass residence times are sufficiently long to cause significant increases in water column salinities that can extend considerably above marine values. Circulation in a lagoon is a composite of gravitational, tidal, and wind-driven components, which all contribute to a typically well-mixed water column, rather than the classic stratified two-layered estuarine regime. Tidal- (e.g., tidal pumping) and wind-driven circulation is particularly pronounced in shallow lagoons that most often occur along low-lying land–sea margins where gravitational circulation is negligible. The development of a hyper-saline water column above freshened submarine groundwater masses can initiate density-driven upward flow. This buoyancy-driven advection/diffusion can enhance the transport of water and its solutes across the sediment-water interface of leaky coastal margins.

1.2   Submarine Groundwater Discharge

The complex interaction of hydrogeologic processes coupled with anthropogenic perturbations within a coastal aquifer control the transport and delivery of subsurface materials as they are exchanged across leaky coastal margins. Recent developments in numerical and mathematical models on the dynamic freshwater–saltwater transition zone serve to better predict future coastal groundwater resources by more quantitatively assessing fresh coastal groundwater reserves as well as the extent and rate of coastal saltwater intrusion. These studies have largely focused on the onshore distribution or trends in groundwater salinities of supply and monitor wells. Attempts to realistically portray and predict the dynamic nature of the freshwater–saltwater transition zone have developed from a need to better constrain the onshore domain of such models by groundwater hydrologists, as well as the need to better understand coastal groundwater characteristics by oceanographers. The focus of this section is on the coastal discharge of groundwater and the implication of this flux to coastal aquifers and ecosystem health, rather than on saltwater intrusion.

While not as evident as surface water runoff, groundwater also flows down gradient and discharges directly into the coastal ocean. The discharge of coastal groundwater has become increasingly important as industry and populations continue to migrate toward fragile coastal zones. The submarine groundwater delivery of certain dissolved constituents such as select radionuclides, trace metals, and nutrient species to coastal bottom waters has often been overlooked [Krest et al., 2000; Valiela et al., 1990; Reay et al., 1992; Simmons, 1992]. This omission from coastal hydrologic and mass balance budgets by both hydrologists and oceanographers alike is largely due to the difficulty in accurately identifying and quantifying submarine groundwater discharge [Burnett et al., 2001a, b; Burnett et al., 2002].

Unfortunately, hydrologists and coastal oceanographers still today sometimes use varied definitions to describe hydrogeologic terms and processes. This problem is clearly manifested in a recent response article by the hydrologist Young [1996] to oceanographer Moore's [1996] very large coastal groundwater flux estimates derived for the mid-Atlantic Bight. There is consequently a real need to merge the disciplines of hydrology and oceanography to develop an integrated approach for studies of coastal

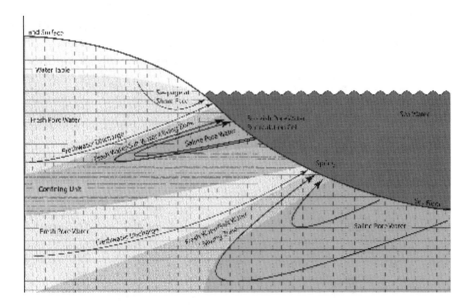

Figure 2: Idealized hydrogeologic description of freshwater/saltwater
exchange processes in a carbonate coastal aquifer.

groundwater discharge [Kooi and Groen, 2001]. In summary, groundwater is
commonly defined simply as water within the saturated zone of geologic
strata [Freeze and Cherry, 1979]. Coastal bottom sediments of an estuary are
obviously saturated, so water within the pores and lattices of submerged
sediments (i.e., pore waters or interstitial waters) can be defined as
groundwater. Therefore, submarine groundwater discharge includes any
upward fluid transfer across the sediment–water interface, regardless of its
age, origin, or salinity. Exchange across this interface is bi-directional
(discharge and recharge), although a net flux is most often upward.

Inland recharge and a favorable underlying geologic framework
control the rate of submarine groundwater discharge within leaky coastal
margins. Figure 2 shows the dominant characteristics of a hypothetical
coastal groundwater system influenced by submarine groundwater discharge.
Freshwater that flows down gradient from the water table toward the sea may
discharge either as diffuse seepage close to shore, or directly into the sea
either as a submarine spring [Swarzenski et al., 2001] or wide scale seepage
[Cable et al., 1999a, b; Corbett et al., 2000a, b, c]. Hydraulic head gradients
that drive freshwater toward the sea can also drive seawater back to the
ocean, creating a saltwater circulation cell. Wherever multiple aquifers and
confining units co-exist, each aquifer will have its own freshwater/saltwater
interface; deeper aquifers will discharge further offshore [Freeze and Cherry,
1979; Bokuniewicz, 1980]. Submarine groundwater discharge is spatially as

well as temporally variable in that both natural and anthropogenic change (i.e., sea-level, tides, precipitation, dredging, groundwater withdrawals) impart a strong signature [Zektzer and Loaiciga, 1993].

Theoretically, submarine groundwater discharge can occur wherever a coastal aquifer is hydrogeologically connected to the sea [Domenico and Schwartz, 1990; Moore and Shaw, 1998; Moore, 1999]. Artesian or pressurized aquifers can extend for considerable distances from shore, and where the confining units are breached or eroded away, groundwater can flow directly into the sea [Manheim and Paull, 1981]. While the magnitude of this submarine groundwater discharge is often less than direct riverine runoff, recent studies have shown that coastal aquifers may contribute significant quantities of freshened water to coastal bottom waters in ideal hydrogeologic strata [Zektzer et al., 1973; Moore, 1996; Burnett et al., 2001a, b; Burnett et al., 2002]. Although it is quite unlikely that submarine groundwater discharge plays a significant role in the global water budget [Zektzer and Loaiciga, 1993], there is strong evidence that suggests that the geochemical signature of many redox sensitive constituents is directly affected by the exchange of subsurface fluids across the sediment–water interface [Johannes, 1980; Giblin and Gaines, 1990; Swarzenski et al., 2001]. This fluid exchange includes direct upward groundwater discharge as well as the reversible exchange at the sediment–water interface (i.e., seawater recirculation) as a result of tidal pumping [Li et al., 1999; Hancock et al., 2000].

1.3 Tools for Submarine Groundwater Discharge

A few methods exist to help identify and quantify submarine groundwater discharge:

1) direct measurement of site-specific exchange (e.g., seepage meters, flux chambers, multi-port samplers),
2) numerical modeling (e.g., MODFLOW, SEAWAT),
3) tracer techniques (e.g., ^{223,224,226,228}Ra, ^{222}Rn, CH_4), and
4) streaming resistivity surveys.

Standard Lee-type seepage meters or more complicated flux chambers have traditionally provided a physical measurement of submarine groundwater discharge across a specific surface area of sediment per unit time. Such physical measurements are time consuming and appear to be most accurate when there is significant upward exchange. There has been considerable advancement in developing a second-generation seep meter, which may either autonomously or manually collect very accurate continuous data on exchange across the sediment–water interface by ultrasound, electromagnetic shifts, or dyes. Even with these advances, such physical measurements are limited to the "foot-print" of the particular device and extrapolations to more regional-scale flux estimates are greatly weakened by the heterogeneous nature of coastal sediments. As a consequence, a precise tracer capable of integrating the spatial heterogeneities of most coastal bottom sediments is needed to derive a realistic estimate of regional exchange. To address this issue, W.S. Moore and W. Burnett and their colleagues (cf. Moore [1996], Moore and Shaw [1998], Moore [1999], Burnett et al. [2001]) have cleverly utilized the four naturally occurring isotopes of radium (^{223,224,226,228}Ra) and ^{222}Rn to study both local and regionally scaled submarine groundwater discharge. Briefly, these radionuclides all are produced naturally in coastal sediments by radioactive decay of their parent isotopes. The half-life of the four Ra isotopes and ^{222}Rn range from about 3.8 days to 1600 years, which coincides ideally with the time frame of many coastal exchange processes. Well-constrained mass balance budgets of these isotopes in coastal waters can therefore provide an estimate of coastal groundwater discharge as well as a means for fingerprinting the various water masses.

While numerical models can range in complexity from simple water balance equations to rigorous variable density transport analysis in heterogeneous media, the inherent assumptions of any model are of course limited in a true portrayal of a particular hydrogeologic regime. That said, models do offer insight in the magnitude or scale of exchange processes and provide a means to evaluate the interdependence of this flux on one or more critical variables. Modeling of coastal groundwater flow has become much more widespread with the availability of PC-based software packages such as MODFLOW, SUTRA, and SEAWAT [McDonald and Harbaugh, 1988; Voss, 1984; Langevin, 2001].

Due to the inherently difficult task of identifying diffuse submarine groundwater discharge from coastal sediments, a tool to rapidly identify sediment pore water conductivities would be very useful. Indeed, F. Manheim and colleagues have successfully adapted a multi-channel horizontal DC streamer array to examine subsurface resistivity anomalies in coastal settings. Such systems, when verified against pore fluid studies and geologic core descriptions, provide unprecedented and highly reliable information on freshened subsurface water masses and the dynamic interplay at the freshwater–saltwater transition zone.

The second section of this chapter will describe two examples of enhanced coastal exchange processes across the sediment–water interface in Florida. Both sites are representative of carbonate platform settings, where various limestone dissolution features can facilitate exchange of coastal groundwater with surface water.

2. CASE STUDY: TAMPA BAY

Tampa Bay ($1,031$ km^2) sits on the central west coast of Florida, and while it has an average depth of only 3.5 m, the navigational channels that extend the full length of the bay reach depths of up to 14 m (Figure 3). Freshwater inputs to the bay include precipitation (roughly 43%), surface water runoff (41%) and smaller contributions from groundwater and industrial/municipal point sources [Zarbock et al., 1995]. Due to the small drainage basin ($6,480$ km^3), the mean (1985–1991) annual surface water runoff rate is less than 100 m^3 sec^{-1} of which about 80% is accounted for by the discharge of four rivers [Zarbock et al., 1995]. Salinities range from seawater values in the lower bay to less than 20 in the upper bays (Hillsborough and Old Tampa), regardless of season. The amount of precipitation as well as climate fluctuations, however, does appear to directly affect the salinity regime of Tampa Bay [Schmidt and Luther, 2002]. Water mass residence times vary considerably (~20–120 days) in the bay, depending on the water depth and riverine input. Any significant coastal groundwater and associated contaminants discharged at sites where the water column is poorly flushed (i.e., long residence times) could deleteriously impact ecosystem health. Streaming resistivity surveys in concert with more detailed pore water geochemistry, geophysics, and geologic descriptions were used to provide information on the geologic control of coastal groundwater aquifers in Tampa Bay. Streaming resistivity data were collected with a positively buoyant 120-m-long streamer cable that consisted of two current electrodes and six receiver dipoles. The electrode resistivities were measured using a high voltage AC-DC converter, a TEM/resistivity transmitter, and a multi-function receiver. Differential GPS navigation, high-resolution bathymetry, and ancillary water column parameters (salinity, conductivity, pH, color, temperature) were also continuously collected and incorporated in the resistivity data stream. Results were initially processed using Zonge TS2DIP inversion software, modeled and then contoured against depth. Figure 4 illustrates an example of a typical pore fluid resistivity cross-section produced during the streaming resistivity survey at a mid-bay site (see Figure 3 for the location in Tampa Bay). Note the elevated apparent resistivities below a depth of about 10 m observed in the uppermost cross section.

A formation factor can provide a site-specific conversion of resistivity to conductivity or salinity. An essential field validation of the streaming resistivity data by down-core pore fluid analysis confirms a dramatic shift in interstitial salinity at a depth of approximately 10 m (Figure 5). From the interpretation of many tens of km of streaming resistivity data

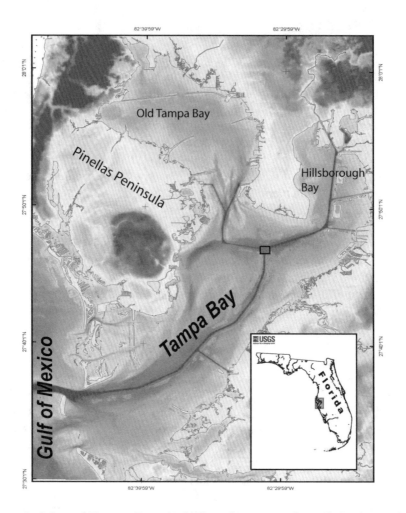

Figure 3: Map of Tampa Bay, including the two major sub-basins and the site (□) of the streaming resistivity survey and deep pore water profile comparison.

in Tampa Bay, it is becoming evident that a large freshened water mass exists in the sediments below about 10 m. How this coastal groundwater migrates through a variably thick and effective confining unit into bay bottom waters is the focus of a larger interdisciplinary effort that ties together a broad range of geologic and hydrologic expertise.

It is likely that these observed freshened water masses beneath Tampa Bay represent paleo-groundwaters, which possibly infiltrated geologic strata during the Pleistocene when sea levels were lower than at

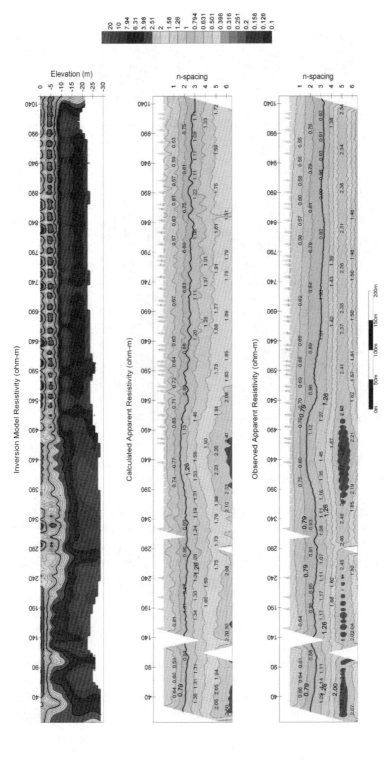

Figure 4: Interpreted cross-section of pore water resistivities at a mid-bay site in Tampa Bay. The bottom graph represents the observed data plotted against the dipole number, the middle graph represents a derived apparent resistivity, and the top graph illustrates the inversion-modeled resistivity relative to depth.

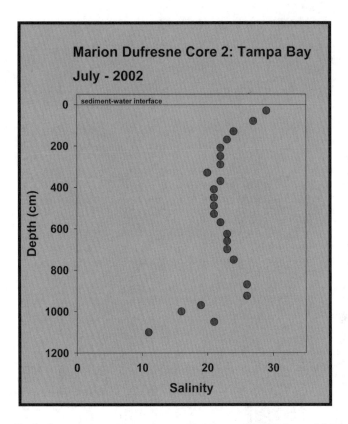

Figure 5: A down-core pore water salinity profile at the mid-bay site. Note the dramatic decline in interstitial salinities at about 10 m depth. These data confirm the observations of the steaming resistivity surveys.

present. Isotopic pore water analysis (i.e., $^{87}Sr/^{86}Sr$) should provide an age constraint to identify the evolution of these subsurface water masses. How quickly the surficial coastal aquifer around the perimeter of Tampa Bay responds to human-induced (i.e., agriculture, industry, groundwater mining) or natural change (i.e., precipitation) is a hydrogeologic question that warrants further investigation in the heavily populated Tampa Bay area. The two modes of submarine groundwater discharge (surficial versus paleo-groundwater) must occur over very different time scales, must release very different waterborne constituents, and most likely utilize different flow regimes during transport. For example, the upward flow of paleo coastal groundwaters may develop through relic dissolution features that are prevalent only in some parts of Tampa Bay, while water within the surficial coastal aquifer could percolate much more rapidly through porous shoreline sands. Because coastal groundwater discharge in Tampa Bay can be divided into these two distinct processes that convey very different water masses and associated constituents, resource and management decisions should account for these variable submarine inputs.

3. CASE STUDY: CRESCENT BEACH SUBMARINE SPRING

Although coastal groundwater can theoretically discharge along any shoreline with a positive potentiometric gradient and favorable underlying geologic framework, this upward flow is most often very diffuse and inherently difficult to identify and quantify. In sharp contrast, offshore groundwater springs do exist, and these sites provide a spectacular opportunity to study the dynamic interface between freshwater and saltwater. Florida has a large number of submarine groundwater springs, which discharge a full range of salinities into coastal waters. Of these, Crescent Beach submarine spring off the northeastern coast of Florida (Figure 6) is among the most distinguished, as it delivers on the order of ~40 m^3 sec^{-1} of salinity 6 water to Atlantic Ocean coastal bottom waters (salinity = 36). Such a large flux of freshened groundwater to the coastal ocean provides a transport mode for land-derived nutrients and other potentially deleterious contaminants such as metals and radionuclides if these groundwaters have a terrigenous origin. This coarse contrast in salinity values of the two mixing water masses can also initiate biogeochemical and physico-chemical reactions that are characteristic of surface-water estuaries. Such reactions can include particle aggregation/coagulation as well as surface complexation reactions that can affect the speciation or behavior of a particular chemical in response to changes in ionic strength.

Through detailed high-resolution seismic surveys and vibra-core descriptions, we have learned that the Miocene-aged confining unit (Hawthorn Group) has been effectively eroded away at Crescent Beach submarine spring. This allows for direct communication of coastal groundwater with Atlantic Ocean bottom waters (Figure 7). Geophysical interpretations also reveal multiple large-scale collapse features directly adjacent to the submarine vent, indicating that the surrounding geologic framework is karst-dominated. This perforated landscape with relict and modern sinkholes and springs is thus a highly effective leaky coastal margin.

In northeastern Florida, water within the highly productive Floridan aquifer system is commonly artesian along the coastal zone. Coastal groundwater is thus under sufficient pressure to flow freely at land surface through limestone conduits, springs, fractures, and other dissolution features. Ocala Limestone groundwater is relatively rich Ca-HCO_3 water that

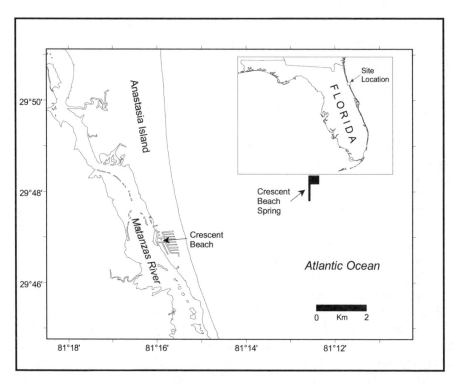

Figure 6: Location map for Crescent Beach submarine spring (Lat 29° 46.087N, Long 81° 12.478W) in NE Florida (adapted from Swarzenski *et al.* [2001]).

generally increases in hardness along a transect from the inland recharge area eastward toward the coast. In coastal northeastern Florida, groundwater chloride concentrations generally increase from north to south and are about 110 mM around the town of Crescent Beach. The geochemical signature of select trace metals and major solutes of Crescent Beach submarine spring water is compared to Atlantic Ocean surface waters in Table 1. CBSS denotes Crescent Beach submarine spring water; the surface seawater site was collected at 1 m depth approximately 100 m from the vent feature. All trace element concentrations are in nM and were measured using a sector field ICP-MS; major solute concentrations are in mM. Expected enrichments were observed in the spring waters for reduced Fe and Mn species, while depletions were noted for the reverse redox couples, U and V. Barium was also elevated considerably in the spring waters, and has recently been suggested as an additional effective coastal groundwater tracer [Shaw *et al.*, 1998].

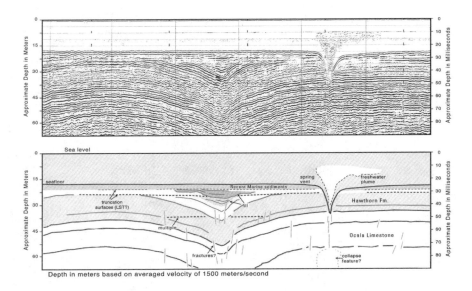

Figure 7: High-resolution seismic interpretation of the geologic framework surrounding the artesian coastal aquifer system at Crescent Beach submarine spring (adapted from Swarzenski *et al.* [2001]).

4. CONCLUSIONS

Leaky coastal margins are defined here as any land–sea margin where the bi-directional exchange of groundwater with seawater is naturally enhanced. Saltwater intrusion into a freshwater coastal aquifer is one well-studied and critical process within leaky coastal margins. This can occur either naturally or where significant groundwater withdrawals have created an artificial low in the potentiometric surface or water table. Another process just as critical to resource managers, groundwater hydrologists, and oceanographers alike is the discharge of coastal groundwater into seawater. This discharge occurs most often as diffuse seepage closest to shore but is typically very difficult to identify and measure. Coastal groundwater discharge may also occur at sites of submarine springs, where vent water mixes directly with ocean water. Coastal groundwater discharge is of interest not only in the accurate quantification of a comprehensive land–sea margin water budget but also in the precise assessment of groundwater-borne nutrient/contaminant loading estimates into coastal waters. Eutrophication and general coastal ecosystem degradation are obvious potential consequences of coastal groundwater discharge.

trace elements	CBSS (nM)	surface seawater (nM)
Mn	90.9	31.0
Mo	9.5	218.7
Ba	300.7	55.3
U	0.1	18.4
V	21.9	47.6
Fe	64.9	3.1
major solutes	(mM)	(mM)
Cl	102.39	545.79
Na	88.74	468.03
SO_4	8.50	28.21
Mg	10.37	53.08
Ca	7.39	10.25
K	1.64	10.21
Sr	0.10	0.09
F	0.04	0.07
Si	0.32	0.18

Table 1: A comparison of select dissolved trace elements (nM) and major solutes (mM) in seawater and Crescent Beach submarine spring water (CBSS).

This paper described hydrogeologic characteristics unique to leaky coastal margins and then illustrated these by providing two examples from Florida. Tampa Bay has both active seepage/spring sites close to shore that respond rapidly to natural/anthropogenic perturbations, as well as large scale freshened water masses (salinity < 10) at depths greater than about 10 m that may leak upward into bay bottom waters. Water budgets in Tampa Bay suggest that submarine groundwater discharge indeed represents a significant component of surface water runoff to the bay. At the Crescent Beach submarine spring site, the upwelling coastal groundwater has a very distinct geochemical signature from that of ambient seawater and presents a direct route of groundwater-borne constituents to the coastal ocean. More information about these two case studies can be found on the accompanying CD.

Recently a multi-disciplinary conference on Leaky Coastal Margins was organized in St. Petersburg, Florida by the U.S. Geological Survey. At this meeting, resources managers and coastal scientists representing varied expertise discussed tools, techniques, and common interests pertaining to leaky coastal margins. More information regarding this meeting can be found at the USGS web site.[1]

Acknowledgments

The following colleagues have provided enjoyable and valuable discussions that have led toward the concept of Leaky Coastal Margins: Jack Kindinger (USGS), Terry Edgar (USGS), Jon Martin (University of Florida), Bill Burnett (Florida State University), Jeff Chanton (Florida State University), Billy Moore (University of Southern California), John Bratton (USGS), Jim Krest (USGS), and Jaye Cable (Louisiana State University). Funding and guidance have been provided largely from the Coastal and Marine Geology Program by John Haines (USGS).

REFERENCES

Beebe, A., "Largest U.S. seawater desalinization plant coming to Tampa Bay," *Water-Engineering Management*, **147**, 8, 2000.

Bokuniewicz, H., "Groundwater seepage into Great South Bay, New York," *Estuarine Coastal Marine Science*, **10**, 437–444, 1980.

Bollinger, M.S. and Moore, W.S., "Evaluation of salt marsh hydrology using radium as a tracer," *Geochimica et Cosmochimica Acta*, **57**, 2203–2212, 1993.

Burnett, W.C., Kim, G., and Lane-Smith, D., "A continuous radon monitor for assessment of radon in coastal ocean waters", *Journal of Radioanalytical and Nuclear Chemistry*, **249**, 167–172, 2001b.

Burnett, W.C., Taniguchi, M., and Oberdorfer, J., "Measurement and significance of the direct discharge of groundwater into the coastal zone," *Journal of Sea Research*, **46/2**, 109–116, 2001a.

Burnett, W.C., Chanton, J., Christoff, J., Kontar, E., Krupa, S., Lambert, M., Moore, W., O'Rourke, D., Paulsen, R., Smith, C., Smith, L., and Taniguchi, M., "Assessing methodologies for measuring groundwater discharge to the ocean," *EOS*, **83**, 117–123, 2002.

Cable, J.E., Bugna, G.C., Burnett, W.C., and Chanton, J.P., "Application of ^{222}Rn and CH_4 for assessment of groundwater discharge to the coastal ocean," *Limnology and Oceanography*, **41**, 1347–1353, 1996a.

---

[1] http://coastal.er.usgs.gov/lc-margins/

Cable, J.E., Burnett, W.C., Chanton, J.P., and Weatherly, G.L, "Estimating groundwater discharge into the northeastern Gulf of Mexico using radon-222," *Earth and Planetary Science Letters*, **144**, 591–604, 1996b.

Corbett, D.R., Dillon, K., and Burnett, W., "Tracing groundwater flow on a barrier island in the northeast Gulf of Mexico," *Estuarine, Coastal, and Shelf Science*, **51**, 227–242, 2000a.

Corbett, D.R., Dillon, K., Burnett, W., and Chanton, J., "Estimating the groundwater contribution into Florida Bay via natural tracers ^{222}Rn and CH_4," *Limnology and Oceanography*, **45**, 1546–1557, 2000b.

Corbett, D.R., Kump, L., Dillon, K., Burnett, W., and Chanton, J., "Fate of wastewater-borne nutrients in the subsurface of the Florida Keys, USA," *Marine Chemistry*, **69**, 99–115, 2000c.

Domenico, P.A. and Schwartz, F.W., "Physical and Chemical Hydrogeology," *John Wiley and Sons, Inc.,* 824 pp., 1990.

Edelman, J.H., "Groundwater hydraulics of extensive aquifers," *International Institute for Land Reclamation and Drainage Bulletin, Wageningen, Netherlands*, **13**, 219 pp., 1972.

Edmond, J.M., Measures, C., McDuff, R.R., Chan, L-H., Collier, R., and Grant, B., "Ridge crest hydrothermal activity and the balance of major and minor elements in the ocean: The Galapagos data," *Earth, Planetary Science Letters*, **46**, 1–18, 1979.

Fanning, K.A., Byrne, R H., Breland, J.A., Betzer, P.H., Moore, W.S., Elsinger, R.J., and Pyle, T.E., "Geothermal springs of the west Florida continental shelf: Evidence for dolomitization and radionuclide enrichment," *Earth and Planetary Science Letters*, **52**, 345–354, 1981.

Freeze, R.A. and Cherry, J.A., *Groundwater*, Prentice-Hall, 1979.

Giblin, A.E. and Gaines, A.G., "Nitrogen inputs to a marine embayment: the importance of groundwater," *Biogeochemistry*, **10**, 309–328, 1990.

Glover, R.E., "The pattern of fresh water flow in a coastal aquifer," *Journal of Geophysical Research*, **64**, 457–459, 1959.

Hancock G.J., Webster, I.T., Ford, P.W., and Moore, W.S., "Using Ra isotopes to examine transport processes controlling benthic fluxes into a shallow estuarine lagoon," *Geochimica et Cosmochimica Acta*, **21**, 3685–3699, 2000.

Henry, H.R., "Interfaces between saltwater and freshwater in coastal aquifers," *USGS Water Supply Paper,* **1613,** C35–C70, 1964.

Johannes, R.E., "The ecological significance of the submarine discharge of groundwater," *Marine Ecology Progress Series*, **3**, 365–373, 1980.

Johnson, T.A. and Whittaker, B., "Saltwater Intrusion in the Coastal Aquifers of Los Angeles County, California," Chapter 2, this volume.

Kohout, F.A., "A hypothesis concerning cycling flow of saltwater related to geothermal heating in the Floridan aquifer," *Transactions of the New York Academy of Sciences*, **28**, 249–271, 1965.

Kooi, H. and Groen J., "Offshore continuation of coastal groundwater systems; predictions using sharp interface approximations and variable-density flow modeling," *Journal of Hydrology*, **246**, 19–35, 2001.

Krest, J.M., Moore W.S., Gardner L.R., and Morris J.T., "Marsh nutrient export supplied by groundwater discharge: Evidence from radium measurements," *Global Biogeochemical Cycles*, **14**, 167–176, 2000.

Langevin, C.D., "Simulation of groundwater discharge to Biscayne Bay, southeastern Florida," *U.S. Geological Survey Water-Resources Investigations Report*, **00-4251**, 127 pp., 2001.

Li, L., Barry, D.A., Stagnitti, F., and Parlange, J.-Y., "Submarine groundwater discharge and associated chemical input to a coastal sea," *Water Resources Research*, **35**, 3253–3259, 1999.

Manheim, F.T. and Paull, C.K., "Patterns of groundwater salinity changes in a deep continental-oceanic transect off the southeastern Atlantic coast of the U.S.A," *Journal of Hydrology*, **54**, 95–105, 1981.

McDonald, M.G. and Harbaugh, A.W., "A modular three-dimensional finite-difference groundwater model," *U.S. Geological Survey Techniques of Water Resources Investigations*, Book 6, 586 pp., 1988.

Meisler, H., Leahy, P.P., and Knobel, L., "Effect of eustatic sealevel changes on saltwater-freshwater in the northern Atlantic coastal plain," *U.S. Geological Survey, Water Supply Paper*, **2255**, 1984.

Moore, W.S. and Shaw T.J., "Chemical signals from submarine fluid advection onto the continental shelf," *Journal of Geophysical Research*, **103**, 21543–21552, 1998.

Moore, W.S., "Large groundwater inputs to coastal waters revealed by [226]Ra enrichments," *Nature*, **380**, 612–614, 1996.

Moore, W.S., "The subterranean estuary: a reaction zone of groundwater and seawater," *Marine Chemistry*, **65**, 111–125, 1999.

Mualem Y. and Bear, J., "The shape of the interface in steady flow in a stratified aquifer," *Water Resource Research,* **10**, 1207–1215, 1974.

Reay, W.G., Gallagher, D.L., and Simmons, G.M., "Groundwater discharge and its impact on surface water quality in a Chesapeake Bay inlet," *Water Resources Bulletin*, **28**, 1121–1134, 1992.

Schmidt, N. and Luther, M.E., "ENSO impacts on salinity in Tampa Bay, Florida," *Estuaries*, **25**, 976–984, 2002.

Shaw, T.J., Moore, W.S., Kloepfer, J., and Sochaski, M.A., "The flux of barium to the coastal waters of the Southeastern United States: the importance of submarine groundwater discharge," *Geochimica et Cosmochimica Acta*, **62**, 3047–3052, 1998.

Simmons, G.M. Jr., "Importance of submarine groundwater discharge (SGWD) and seawater cycling to material flux across sediment/ water interfaces in marine environments," *Marine Ecology Progress Series*, **84**, 173–184, 1992.

Swarzenski, P.W., Reich, C.D., Spechler, R.M., Kindinger, J.L., and Moore, W.S., "Using multiple geochemical tracers to characterize the hydrogeology of the submarine spring off Crescent Beach, Florida," *Chemical Geology*, **179**, 187–202, 2001.

Thomas, H.E., "Groundwater regions of the United States—their storage facilities," U.S. 83rd Congress, House Interior and Insular affairs Committee, *The Physical and Economic Foundation of Natural Resources*, **3**, 78 pp., 1952.

Valiela, I., Costa, J., Foreman, K., Teal, J.M., Howes, B., and Aubrey, D., "Transport of groundwater-borne nutrients from watersheds and their effects on coastal waters," *Biogeochemistry*, **10**, 177–197, 1990.

Vidal, V.M.V., Vidal, F.V., Isaacs, J.D., and Young, D.R., "Coastal submarine hydrothermal activity off northern Baja California," *Journal of Geophysical Research*, **83**, 1757–1774, 1978.

Voss, C.I., "SUTRA: A finite-element simulation model for saturated— unsaturated fluid-density-dependent groundwater flow with energy transport or chemically reactive single species solute transport," *U.S. Geological Survey Water-Resources Investigation Report* 84-4369, 409 pp., 1984.

Young, P.L., "Submarine groundwater discharge," *Nature*, **382**, 121–122, 1996.

Zektzer, I.S. and Loaiciga, H., "Groundwater fluxes in the global hydrological cycle: Past, present and future," *Journal of Hydrology*, **144**, 405–427, 1993.

Zektzer, I.S., Ivanov, V.A., and Meskheteli, A.V., "The problem of direct groundwater discharge to the seas," *Journal of Hydrology*, **20**, 1–36, 1973.

Zarback, H., Janicki, A., Wade, D., Heimbuch, D., and Wilson, H., "Current and historical freshwater inflows to Tampa Bay," *Tampa Bay National Estuary Program*, St. Petersburg, FL, 1995.

CHAPTER 6

Tidal Dynamics of Groundwater Flow and Contaminant Transport in Coastal Aquifers

L. Li, D.A. Barry, D.-S. Jeng, H. Prommer

1. INTRODUCTION

Coastal/estuarine water pollution is becoming an increasingly serious global problem largely due to input of land-derived contaminants. For example, nutrient leachate from the sugar cane production areas of North-East Queensland is causing great concern for the Great Barrier Reef in Australia [Haynes and Michael-Wagner, 2000]. The resulting degradation of coastal resources affects significantly economic and social developments of coastal regions. Traditionally, terrestrial fluxes of chemicals to coastal water have been estimated on the basis of river flow alone. However, recent field observations indicate that contaminants entering coastal seas and estuaries with groundwater discharge (submarine groundwater discharge, SGWD) can significantly contribute to coastal pollution, especially in areas where serious groundwater contamination has occurred (e.g., Moore [1996], Burnett *et al.* [2001]). The International Geosphere-Biosphere Programme (IGBP) [Buddemeier, 1996] has identified submarine groundwater discharge as an important but rather unknown source of contamination for coastal marine and estuarine environments. As the groundwater contamination problem worsens, the SGWD may become a dominant source of coastal pollution in certain areas.

SGWD consists of both groundwater flow from upland regions and water exchange at the aquifer–ocean interface [Simmons, 1992]. While the upland groundwater flow can be estimated based on the aquifer recharge [Zekster and Loaiciga, 1993], it is difficult to quantify the rate of water exchange across the seabed, which is influenced by near-shore processes [Li *et al.*, 1997a; Turner *et al.*, 1997; Li and Barry, 2000]. Large rates of SGWD, derived from geochemical signals of enriched natural tracers (e.g., ^{226}Ra) [Moore, 1996] in coastal seas, have been found excessive and cannot

1-56670-605-X/04/$0.00+$1.50
© 2004 by CRC Press LLC

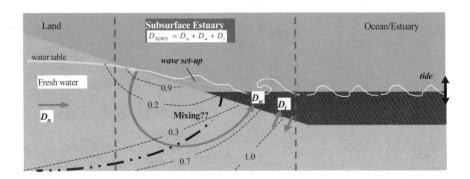

Figure 1: A simple model of SGWD consisting of inland fresh groundwater flow (D_n) and seawater recycling (water exchange due to wave set-up D_w, and due to tides D_t). The mixing of the recycling water with fresh groundwater results in the near-shore salinity profiles as schematically shown by the thin dashed lines (in contrast with the traditional saltwater wedge view shown by the thick dot-dashed line).

be supported by the aquifer recharge [Younger, 1996]. This suggests that water exchange at the interface may have constituted a large portion of the SGWD. A theoretical model of SGWD has been developed to include tidally oscillating groundwater flow and circulation due to wave set-up (i.e., on-shore tilt of the mean sea level; Figure 1). These two local processes were found to cause a large amount of water exchange across the interface.

Although the exchanging/recycling water is largely of marine origin, it mixes and reacts with groundwater and aquifer sediments, modifying the composition of the discharging water. The exchange processes can reduce the residence time of chemicals in the mixing zone of the aquifer, similar to tidal flushing of a surface estuary [Li et al., 1999]. As a result, the rates of chemical fluxes from the aquifer to the ocean increase but the exit chemical concentrations are reduced (dilution effects). The exchange can also alter the geochemical conditions (redox state) in the aquifer and affect the chemical reactions. It has been shown numerically that the exchange enhances the mixing of oxygen-rich seawater and groundwater, and creates an active zone for aerobic bacterial populations in the near-shore aquifer. This zone leads to a considerable reduction in breakthrough concentrations of aerobic biodegradable contaminants at the aquifer–ocean interface [Enot et al., 2001; Li et al., 2001].

In essence, the water exchange and subsequent mixing of the recycling water with fresh groundwater, driven by the oceanic oscillations, lead to the creation of subsurface estuary (subsurface analogue to surface estuary) as suggested by Moore [1999]. The role of a subsurface estuary in determining the terrestrial chemical input to the sea may be compared with that of a surface estuary. Most previous studies of coastal groundwater, focusing on large-scale saltwater intrusion in aquifers, have ignored the dynamic effects of tides and waves on the flow and mixing processes in the near-shore area of the aquifer (e.g., Huyakorn *et al.* [1987]). Despite some early work on coastal groundwater flow and discharge to the sea [Cooper, 1959], it was not until the 1980s that researchers began to investigate the environmental and ecological impacts of groundwater discharge [Bokuniewicz, 1980; Johannes, 1980].

Globally, the fresh groundwater discharge has been estimated to be a few percent of the total freshwater discharge to the ocean [Zekster and Loaiciga, 1993]. Recently, Moore [1996] conducted experiments on [226]Radium enrichment in the coastal sea of the South Atlantic Bight. From the measurements, he inferred, on the basis of mass balance, that groundwater discharge amounts to as much as 40% of the total river flow into the ocean in the study area. This estimate contrasts with previous figures that range from 0.1 to 10%. Younger [1996] suggested that the recharge to the coastal aquifer could only support 4% of the estimated discharge. A model that includes recycling/exchanging water across the seabed was found to predict the excessive discharge rate [Li *et al.*, 1999].

Since the exchanging water is largely of marine origin, its impact on the fate of chemicals in the aquifer and chemical fluxes to coastal water depends on its mixing with groundwater. Laboratory experiments have revealed large tide-induced variations of flow velocities and salinity in the intertidal zone of the aquifer. This suggests that the mass transport of salinity is affected by tides significantly. The mixing of the tide-induced recycling water with fresh groundwater results in a salinity profile of two saline plumes near the shore [Boufadel, 2000], as schematically shown in Figure 1. The mixing zone is in contrast with the traditional saltwater wedge. Field measurements also showed fluctuations of salinity near the shore in response to tides and waves [Nielsen, 1999; Cartwright and Nielsen, 2001]. Tidal effects on salinity distribution in the aquifer have also been demonstrated by numerical investigations [Ataie-Ashtiani *et al.*, 1999; Zhang *et al.*, 2001]. These results suggest the existence of a mixing zone of the coastal aquifer that behaves much like an estuary. In this subsurface estuary, flow and mass transport/transformation are affected by both the net groundwater flow and water exchange/mixing induced by oceanic oscillations, particularly the tides.

The first part of the chapter is on the effects of tides on coastal groundwater, focusing on the water table fluctuations in shallow aquifers. Various analytical solutions of the tide-induced groundwater table fluctuations under different conditions will be presented briefly. The second

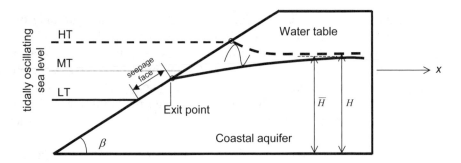

Figure 2: Schematic diagram of tidal conditions at the beach face and water table fluctuations in an unconfined aquifer.

part of the chapter is to examine the effects of the tide-induced groundwater fluctuations on the fate of chemicals in the near-shore aquifer and chemical fluxes to coastal waters. The discussion is based on several on-going studies aiming to improve the understanding and quantification of subsurface pathways and fluxes of chemicals to coastal environments. Additional materials of animated numerical simulation results and color plots are available on the accompanying CD.

2. TIDE-INDUCED GROUNDWATER OSCILLATIONS IN COASTAL AQUIFERS

Groundwater heads in coastal aquifers fluctuate in responses to oceanic tides. Such fluctuations have been subject to numerous recent studies (e.g., Nielsen [1990], Turner [1993], Li et al. [1997a], Nielsen et al. [1997], Baird et al. [1998], Raubenheimer et al. [1999], Li et al. [2000], Li and Jiao [2002a, b], Jeng et al. [2002]). In unconfined aquifers, such responses are manifested as water table fluctuations. These fluctuations are attenuated as they propagate inland, while the phases of the oscillations are shifted [Nielsen, 1990]. Modeling of tidal groundwater head fluctuations are often based on the Boussinesq equation assuming negligible vertical flow,

$$\frac{\partial h}{\partial t} = D \frac{\partial^2 h}{\partial x^2} \qquad (1)$$

where h is the groundwater head fluctuation ($H - \overline{H}$, H is the total head and $\overline{H}$ is the mean head) as shown in Figure 2; x is the inland distance from the shore; t is time; and D is the hydraulic diffusivity, $= T/S$ (S and T are the aquifer's storativity/specific yield and transmissivity, respectively). Note that Eq. (1) is a linearized Boussinesq equation. Although it is applicable to both confined and unconfined aquifers, the application to the latter requires that the tidal amplitude be relatively small with respect to the mean aquifer thickness [Parlange *et al.*, 1984]. The effects of nonlinearity will be discussed later.

Analytical solutions for predicting tidal groundwater head fluctuations are available, for example [Nielsen, 1990],

$$h = A_0 \exp(-\kappa x)\cos(\omega t - \kappa x) \tag{2a}$$

where A_0 and ω are the tidal amplitude and frequency, respectively; κ is the rate of amplitude damping and phase shift, and is related to the tidal frequency and the aquifer's hydraulic diffusivity,

$$\kappa = \sqrt{\frac{\omega}{2D}} \tag{2b}$$

The solution can be presented in an alternative form,

$$h = \mathrm{Re}\big[A_0 \exp(i\omega t - kx)\big] \tag{3a}$$

with

$$\kappa = \sqrt{\frac{i\omega}{D}} \tag{3b}$$

where $i = \sqrt{-1}$ and k is the complex wave number. The relation expressed by Eq. (3b) is termed wave dispersion. The solution assumes that the seaward boundary condition of the groundwater head is defined by the tidal sea level oscillations, i.e.,

$$h(0,t) = A_0 \cos(\omega t) \tag{4}$$

Far inland ($x \to \infty$), the gradient of h is taken to be zero (the tidal effects are diminished), i.e.,

$$\left.\frac{\partial h}{\partial x}\right|_{x=+\infty} = 0 \tag{5}$$

This simple solution also assumes: $A_0/\bar{H}$ (for unconfined aquifers) small, i.e., negligible nonlinear effects; vertical beach face; negligible capillary effects; no leakage exchange between shallow and deep aquifers; negligible vertical flow effects ($\bar{H}\kappa$ small); and negligible seepage face formation. In the following, we discuss relevant effects in situations where these assumptions do not hold.

2.1   Nonlinear Effects

Parlange *et al.* [1984] examined the nonlinear effects. Their analysis is briefly presented here. The head fluctuation is governed by the nonlinear Boussinesq equation as follows,

$$\frac{\partial h}{\partial t} = \frac{K}{n_e} \frac{\partial}{\partial x}\left[\left(\bar{H} + h\right)\frac{\partial h}{\partial x}\right] \tag{6}$$

The seaward and landward boundary conditions are the same as described by Eqs. (4) and (5). A perturbation technique is applied to solve Eq. (6). The solution of h is sought for in the following form,

$$h = \bar{H}\left[\varepsilon h_1 + \varepsilon^2 h_2 + O\left(\varepsilon^3\right)\right] \tag{7}$$

where ε is the perturbation variable, $= A_0/\bar{H}$; it is less than unity under normal conditions. Substituting Eq. (7) into Eq. (6) results in two perturbation equations for h_1 and h_2. Solving these two equations gives,

$$h = A_0 \exp\left(-\kappa x\right)\cos\left(\omega t - \kappa x\right) +$$

$$\frac{A_0^2}{2\bar{H}}\left[\exp\left(-\sqrt{2}\kappa x\right)\cos\left(2\omega t - \sqrt{2}\kappa x\right) - \exp\left(-2\kappa x\right)\cos\left(2\omega t - 2\kappa x\right)\right] \tag{8}$$

$$+\frac{A_0^2}{4\bar{H}}\left[1 - \exp\left(-2\kappa x\right)\right]$$

The nonlinear effects as shown by Eq. (8) lead to the generation of a second harmonics (the second term of the right-hand side with frequency 2ω) and a water table overheight (increase of the mean water table height; the third term of the RHS). The superposition of the second harmonics and the primary signal gives rise to the asymmetry between the rising and falling phases of the water table fluctuations, often observed in the field.

2.2   Slope Effects

Nielsen [1990] reported the first analytical investigation on the slope effects. He derived a perturbation solution for small amplitude water table fluctuations based on the linearized Boussinesq equation by matching a prescribed series solution with the moving boundary condition due to the slope. Later, Li *et al.* [2000] presented an improved approach, as described below. To focus on the slope effects, only small amplitude tides are considered, as modeled by the linearized Boussinesq equation (1) subject to the boundary conditions defined by Eqs. (5) and (9).

As shown in Figure 3, tidal oscillations on a sloping beach create a moving boundary:

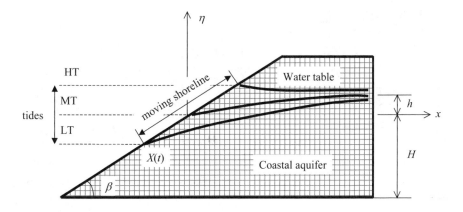

Figure 3: Schematic diagram of water table fluctuations in a coastal aquifer
subject to tidal oscillations at the sloping beach face.

$$h\big[X(t),t\big]=\eta(t) \quad \text{and} \quad X(t)=\cot(\beta)\eta(t) \tag{9}$$

where $X(t)$ is the x-coordinate of the moving boundary (the origin of the x-
coordinate is located at the intersection between the mid-tidal sea level and
the beach face), β is the beach angle, and $\eta(t)$ represents tide-induced
oscillations of the mean sea level. By introducing a new variable
$z = x - X(t)$, Eqs. (1), (9) and (5) are, respectively, transformed to,

$$\frac{\partial h}{\partial t} = D\frac{\partial^2 h}{\partial z^2} - v(t)\frac{\partial h}{\partial z} \tag{10a}$$

$$h(0,t) = \eta(t) \tag{10b}$$

$$\left.\frac{\partial h}{\partial z}\right|_{z=+\infty} = 0 \tag{10c}$$

where

$$v(t) = -\frac{dX(t)}{dt} = A\omega\cot(\beta)\sin(\omega t) \tag{10d}$$

The moving boundary problem of Eq. (1) is thus mapped to a fixed boundary
problem of Eq. (10). A perturbation approach is adopted to solve Eq. (10),
i.e.,

$$h = h_0 + \varepsilon h_1 + O(\varepsilon^2) \tag{11}$$

where $\varepsilon = A\,\kappa\cot(\beta)$ and κ is given by Eq. (2b). The solution is,

$$h = A\exp(-\kappa z)\cos(\omega t - \kappa z) +$$

$$\frac{A\varepsilon}{2}\left\{1 + \sqrt{2}\exp(-\sqrt{2}\kappa z)\cos\left(2\omega t - \sqrt{2}\kappa z + \frac{\pi}{4}\right) - \right. \quad (12)$$

$$\left. \sqrt{2}\exp(-\kappa z)\left[\cos\left(2\omega t - \kappa z + \frac{\pi}{4}\right) + \cos\left(\kappa z - \frac{\pi}{4}\right)\right]\right\} + O(\varepsilon^2)$$

To obtain the solution in the x-coordinate, one can substitute $z = x - A\cos(\beta)\cos(\omega t)$ into Eq. (12). The solution shows that the slope effects are qualitatively similar to those caused by the nonlinearity of finite amplitude tides, i.e., generation of the sub-harmonics and water table overheight.

2.3   Capillary Effects

Parlange and Brutsaert [1987] derived a modified Boussinesq equation to include the capillary effects,

$$\frac{\partial h}{\partial t} = \frac{K\bar{H}}{n_e}\frac{\partial^2 h}{\partial x^2} + \frac{B\bar{H}}{n_e}\frac{\partial^3 h}{\partial t\,\partial x^2} \quad (13)$$

where B is the average depth of water held in the capillary zone above the water table. Barry et al. [1996] solved this equation subject to the boundary conditions described by Eqs. (4) and (5),

$$h = A_0\exp(-\kappa_1 x)\cos(\omega t - \kappa_2 x) \quad (14a)$$

with

$$\kappa_1 = \sqrt{\frac{n_e\omega}{2\bar{H}}\left[\frac{1}{\sqrt{K^2 + (\omega B)^2}} + \frac{\omega B}{K^2 + (\omega B)^2}\right]} \quad (14b)$$

and

$$\kappa_2 = \sqrt{\frac{n_e\omega}{2\bar{H}}\left[\frac{1}{\sqrt{K^2 + (\omega B)^2}} - \frac{\omega B}{K^2 + (\omega B)^2}\right]} \quad (14c)$$

The capillary effects cause the difference between the damping rate (κ_1) and the wave number (κ_2). The solution suggests that capillary effects are only important for high frequency oscillations. Under normal conditions, the effects of unsaturated flows on the tidal water table fluctuations are small. More detailed discussion on the capillary effects can be found in Li et al. [1997b].

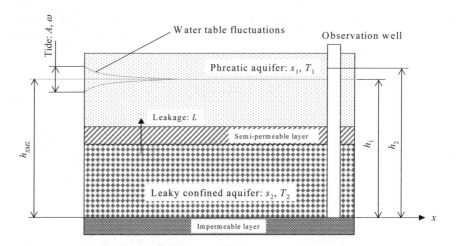

Figure 4: Schematic diagram of a leaky confined aquifer with an overlying phreatic aquifer.

2.4   Leakage Effects

In the above solutions, the bottom boundary of the aquifer is assumed to be impermeable. In reality, it is not uncommon to find composite aquifer systems such as the one shown in Figure 4: an unconfined aquifer overlying and separated from a confined aquifer by a thin semi-permeable layer. The groundwater heads fluctuate in both the confined and the phreatic aquifer. The two aquifers interact with each other via leakage through the semi-permeable layer [e.g., Bear, 1972]:

$$s_1 \frac{\partial h_1}{\partial t} = T_1 \frac{\partial^2 h_1}{\partial x^2} + L(h_2 - h_1) \qquad (15a)$$

$$s_2 \frac{\partial h_2}{\partial t} = T_2 \frac{\partial^2 h_2}{\partial x^2} + L(h_1 - h_2) \qquad (15b)$$

where h_1 and h_2 are the heads in the confined and the phreatic aquifers, respectively; T_1 and T_2 are the transmissivities of these two aquifers, respectively; s_1 is the specific yield of the phreatic aquifer and s_2 is the storativity of the confined aquifer; and L is the specific leakage of the semi-permeable layer.

In reality, the damping of the tidal signal in the unconfined aquifer is much higher than that in the confined aquifer (since $s_1 \gg s_2$). Usually the fluctuations of h_1 become negligible 100 m landward of the shoreline while the tides propagate much further inland in the confined aquifer. Jiao and Tang [1999] solved Eq. (15b) assuming that h_1 is constant, i.e., neglecting the tidal fluctuations in the unconfined aquifer. Their solution shown below suggests that the leakage reduces the tidal signal in the confined aquifer significantly, i.e., the damping rate increases:

$$h_2 = h_{MSL} + A_0 \exp(-k_{L1}x)\cos(\omega t - k_{L2}x)$$ (16a)

with

$$\kappa_{L1} = \sqrt{\frac{L}{2T_2} + \frac{1}{2}\sqrt{\left(\frac{\omega s_2}{T_2}\right)^2 + \left(\frac{L}{T}\right)^2}}$$ (16b)

and

$$\kappa_{L2} = \frac{\omega s_2}{2T_2 \kappa_{L1}}$$ (16c)

Jeng *et al.* [2002] solved the coupled equations (15a) and (15b). Their solution also demonstrates the reduction of tidal signal in the confined aquifer due to leakage. However, the extent of the reduction is less than predicted by Eq. (16). The solution also indicates that the water table fluctuation in the unconfined aquifer is enhanced as a result of the leakage.

2.5 Low Frequency Oscillations

The above solutions consider only one tidal constituent. In reality, tides are more complicated and often bichromatic, containing oscillations of two slightly different frequencies: semi-diurnal solar tide with period $T_1 = 12$ h and frequency $\omega_1 = 0.5236$ Rad h^{-1}, and semi-diurnal lunar tide with $T_2 = 12.42$ h and $\omega_2 = 0.5059$ Rad h^{-1}. As a result, the spring-neap cycle (i.e., the tidal envelope) is formed with a longer period, $T_{sn} = 2\pi/(\omega_1 - \omega_2) = 14.78$ d. Raubenheimer *et al.* [1999] observed water table fluctuations of period T_{sn}. These fluctuations (called spring-neap tidal water table fluctuations, SNWTF) occurred much further inland than the primary tidal signals (diurnal and semi-diurnal tides). While one may relate this long period fluctuation to the spring-neap cycle, the cause of such a phenomenon is not readily apparent. Spring-neap tides are bichromatic signals as described by

$$\eta(t) = A_1 \cos(\omega_1 t) + A_2 \cos(\omega_2 t - \delta)$$ (17)

where A_1 and A_2 are the amplitude of the semi-diurnal solar and lunar tide, respectively, and δ is the phase difference between them. Only two primary forcing signals exist at the boundary. If they propagate in the aquifer independently (as would occur in a linearized model assuming a vertical beach face), the water table response will also be bichromatic and simply described by $A_1 \exp(-\kappa_1 x)\cos(\omega_1 t - \kappa_1 x) + A_2 \exp(-\kappa_2 x)\cos(\omega_2 t - \delta - \kappa_2 x)$.

Both κ_1 ($\sqrt{\dfrac{n_e \omega_1}{2KH}}$) and κ_2 ($\sqrt{\dfrac{n_e \omega_2}{2KH}}$) are high damping rates corresponding to the semi-diurnal frequencies. A slowly damped spring-neap tidal water table fluctuation is not predicted. However, the beach face is sloping and creates a moving boundary as discussed in Section 2.2. The moving boundary induces interactions between the two primary tidal signals as they propagate inland. Such interactions lead to the generation of the SNWTF.

Li et al. [2000] reported an analytical study on the SNWTF. The same approach as described in Section 2.2 was adopted to solve the Boussinesq equation subject to the bichromatic tides. The solution is as follows,

$$h = h_0 + \left(h_{10} + h_{11} + h_{12} + h_{13} + h_{14} \right)\varepsilon + O\left(\varepsilon^2\right) \tag{18a}$$

$$h_0 = A_1 \exp\left(-\kappa_1 x\right)\cos\left(\omega_1 t - \kappa_1 x\right) + A_2 \exp\left(-\kappa_2 x\right)\cos\left(\omega_2 t - \kappa_2 x - \delta\right) \tag{18b}$$

$$h_{10} = \frac{1}{2}\left(A_1 + r_2 \sqrt{r_1} A_2 \right) \tag{18c}$$

$$h_{11} = \frac{\sqrt{2} A_1}{2}\exp\left(-\sqrt{2}\kappa_1 x\right)\cos\left(2\omega_1 t - \sqrt{2}\kappa_1 x + \frac{\pi}{4} \right) \tag{18d}$$

$$h_{12} = \frac{\sqrt{2} r_2 \sqrt{r_1} A_2}{2}\exp\left(-\sqrt{2}\kappa_2 x\right)\cos\left(2\omega_2 t - \delta - \sqrt{2}\kappa_2 x + \frac{\pi}{4} \right) \tag{18e}$$

$$h_{13} = \frac{\sqrt{2}\left(1 + \sqrt{r_1}\right) A_2}{2}\exp\left(-\kappa_3 x\right)\cos\left(\omega_3 t - \delta - \kappa_3 x + \frac{\pi}{4} \right) \tag{18f}$$

$$h_{14} = \frac{\sqrt{2 r_1 + 2} A_2}{2}\exp\left(-\kappa_4 x\right)\cos\left(\omega_4 t - \delta - \kappa_4 x + \theta\right) \tag{18g}$$

where $\varepsilon = A_1 \kappa_1 \cot(\beta)$, $r_1 = \omega_2 / \omega_1 = 0.9662$, $r_2 = A_2 / A_1$, $\omega_3 = \omega_1 + \omega_2$, $\kappa_3 = \sqrt{1 + r_1}\kappa_1$, $\omega_4 = \omega_1 - \omega_2$, $\kappa_4 = \sqrt{1 - r_1}\kappa_1$ and $\theta = \arctan\left(\dfrac{1 - \sqrt{r_1}}{1 + \sqrt{r_1}}\right)$.

The solution indicates that, in a bichromatic tidal system, the moving boundary condition generates an overheight ($h_{10}\varepsilon$), and additional harmonic waves of frequency $2\omega_1$ ($h_{11}\varepsilon$), $2\omega_2$ ($h_{12}\varepsilon$), $\omega_1 + \omega_2$ ($h_{13}\varepsilon$) and $\omega_1 - \omega_2$ ($h_{14}\varepsilon$). The oscillation of $\omega_1 - \omega_2$ represents the spring-neap tidal water table fluctuations. Since the damping rate, κ_4, is much smaller than κ_1, κ_2, and κ_3, the SNWTF propagates much further inland, with a damping distance $(1/\kappa_4)$ five times larger than those for the primary mode water table fluctuations.

2.6   Vertical Flow Effects (Intermediate Depth)

The validity of the Boussinesq equation depends on the shallowness of the aquifer, i.e., $n_e \omega \bar{H}/K$ small [Parlange et al., 1984; Nielsen et al., 1997]. For aquifers of intermediate depths, the vertical flow effects become considerable, in which case the Boussinesq equation needs to be expanded to include high-order terms, e.g., [Parlange et al., 1984]

$$\frac{\partial h}{\partial t} = \frac{K}{n_e} \frac{\partial}{\partial x}\left[h\frac{\partial h}{\partial x} + h^2 \frac{\partial h}{\partial x}\frac{\partial^3 h}{\partial x^2} + \frac{1}{3}h^3\frac{\partial^3 h}{\partial x^3}\right] \tag{19}$$

In a linearized form,

$$\frac{\partial h}{\partial t} = \frac{K\bar{H}}{n_e}\left[\frac{\partial^2 h}{\partial x^2} + \frac{\bar{H}^2}{3}\frac{\partial^4 h}{\partial x^4}\right] \tag{20}$$

The solution to Eq. (20) subject to the usual tidal boundary conditions is

$$h = A_0 \exp\left(-\kappa_{v1}x\right)\cos\left(\omega t - \kappa_{v2}x\right) \tag{21a}$$

with

$$\kappa_{v1} = \mathrm{Re}\left(\frac{1}{\bar{H}}\sqrt{\frac{3}{2}}\sqrt{-1+\sqrt{1+\frac{4\, i\omega n_e \bar{H}}{3}\frac{}{K}}}\right) \tag{21b}$$

$$\kappa_{v2} = \mathrm{Im}\left(\frac{1}{\bar{H}}\sqrt{\frac{3}{2}}\sqrt{-1+\sqrt{1+\frac{4\, i\omega n_e \bar{H}}{3}\frac{}{K}}}\right) \tag{21c}$$

The behavior of κ_{v1} and κ_{v2} is different from that predicted by the Boussinesq solution. The vertical flow effects lead to difference between the damping rate and wave number (the rate of phase shift). In particular, the signal appears to propagate faster than predicted by the Boussinesq solution, i.e., smaller phase shifts.

Using a Rayleigh expansion of the hydraulic potential function in terms of the aquifer depth, Nielsen et al. [1997] derived a groundwater oscillation equation that includes an infinite number of high-order terms to account for the vertical flow effects.

2.7   Density Effects

The above solutions ignore the density effects due to seawater intrusion in the aquifer. Wang and Tsay [2001] investigated the density effects based on a sharp interface approach and derived a governing equation for h including the density effects,

$$\frac{\partial h}{\partial t} = \frac{K}{n_e} \frac{\partial}{\partial x}\left[\left(\delta\eta + H\right)\frac{\partial h}{\partial x}\right] \tag{22}$$

where η is the height of the saltwater–freshwater interface from the base of the aquifer (H is the height of the water table also from the base of the aquifer) and δ is given by

$$\delta = \frac{\rho_s v_f}{\rho_f v_s} - 1 \tag{23}$$

where ρ_f and ρ_s are the density of the freshwater and seawater, respectively; and v_f and v_s are the kinematic viscosity of freshwater and seawater, respectively. Taking $\rho_f = 1000$ kg/m^3, $\rho_s = 1020$ kg/m^3, $v_f = 1.01 \times 10^{-6}$ m^2/s, and $v_s = 1.06 \times 10^{-6}$ m^2/s, δ is calculated to be –0.028. The ratio of $\delta\eta$ to H is at the maximum (near the shoreline where η is at the maximum, being close to H) –0.028. Equation (22) can therefore be approximated by the Boussinesq equation. In other words, the density effects on the water table fluctuations are negligible.

2.8   Seepage Face Effects

In reality, the occurrence of seepage faces is commonplace, in which case the exit point of the water table at the beach face is decoupled from the tidal signal (Figure 2). The boundary condition is then defined by the movement of the exit point rather than the tidal level. Based on the following model of Dracos [1963] and Turner [1993], one can show that the formation of seepage faces reduces the primary forcing signals (semi-diurnal solar and lunar tides) and causes a spring-neap forcing oscillation on the boundary. The inland propagation of this oscillation leads to the SNWTF too.

In the Turner/Dracos model, the movement of the exit point is described by,
Coupling phase:

$$z_e = z_s \quad \text{for } V_{tide} \geq -\frac{K}{n_e}\sin^2\left(\beta\right) \tag{24a}$$

Decoupling phase:

$$z_e = z_{ep} - \frac{K}{n_e}\sin^2\left(\beta\right)\left(t - t_{ep}\right) \quad \text{for } V_{tide} < -\frac{K}{n_e}\sin^2\left(\beta\right) \tag{24b}$$

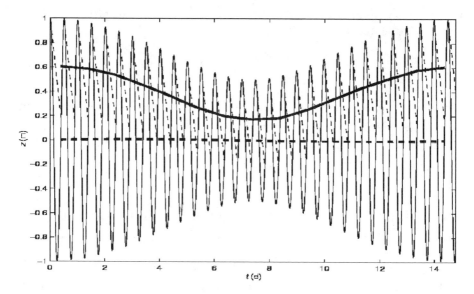

Figure 5: Calculated elevations of the sea level (thin solid line) and the exit point (thin dashed line). Thicker solid and dashed lines show the 25-h averaged elevations of the sea level and the exit point, respectively.

where z_e and z_s are the elevations of the exit point and shoreline, respectively; V_{tide} is the tidal velocity; t_{ep} is the instant when decoupling commences; and z_{ep} is the elevation of the exit point at time t_{ep}.

As an example, Figure 5 shows the calculated seepage face over a spring-neap cycle using the above model. The long period (of T_{sn}) oscillation is clearly evident in the exit point's movement. Further analysis based on the Fourier transformation shows that large oscillations occur at the spring-neap frequency while the amplitudes of the semi-diurnal oscillations are reduced by a factor of 0.4.

3. IMPLICATIONS FOR CONTAMINANT TRANSPORT AND TRANSFORMATION IN TIDALLY INFLUENCED COASTAL AQUIFERS

As demonstrated above, the tides affect significantly the coastal groundwater. The water table fluctuations are the manifestation of such effects in the shallow unconfined aquifer and have been studied extensively. These fluctuations result in oscillating groundwater flow in the near-shore area of the aquifer, enhancing the water exchange and mixing between the aquifer and coastal sea/estuary. In the following, we illustrate the importance

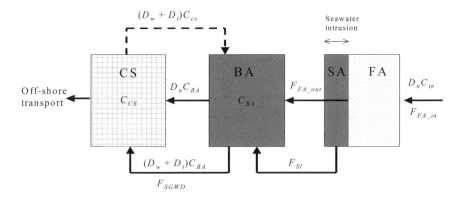

Figure 6: A box model of chemical transfer from the aquifer to coastal sea.

of these local flow, exchange, and mixing processes for chemical transport and transformation in the near-shore aquifer and the associated chemical fluxes to coastal water.

3.1 Tide-Induced Flushing and Dilution Effects on Chemical Transport Processes

Li *et al.* [1999] developed a model of SGWD that incorporates the outflows of the tide-induced oscillating groundwater flow and wave-induced groundwater circulation as well as the net groundwater discharge (Figure 1),

$$D_{SGWD} = D_n + D_w + D_t \qquad (25)$$

Using a "box" model described below, Li *et al.* [1999] examined the importance of SGWD, especially D_w and D_t, on the process of chemical transfers from the aquifer to the ocean.

The model includes three water bodies: coastal sea (CS), brackish aquifer (BA), and freshwater aquifer (FA). Chemical transfers occur between the water bodies as shown by arrows in Figure 6. The chemicals are assumed to be strongly absorbed by sand particles in fresh groundwater and to desorb into brackish groundwater. The mass balance for FA can be described by

$$F_{FA\_in} = F_{FA\_out} \quad \text{if} \quad S = S_{eq} \qquad (26a)$$

$$F_{FA\_out} = 0 \quad \text{and} \quad V_{FA}\frac{dS}{dt} = F_{FA\_in} \quad \text{if} \quad S < S_{eq}, \qquad (26b)$$

$$F_{FA\_in} = D_n C_{in} \qquad (26c)$$

$$S_{eq} = K_d C_{in} \qquad (26d)$$

where F_{FA_in} and F_{FA_out} are the input and output mass flux for FA, respectively; S is the amount of absorbed chemical and the subscript eq denotes the equilibrium state; V_{FA} is the effective volume of the FA; K_d is the distribution coefficient; and C_{in} is the input chemical concentration. Equations (26a) and (26b) express the equilibrium and non-equilibrium states, respectively.

For BA, the governing equations are:

$$V_{BA}\frac{dC_{BA}}{dt} = \left(F_{FA\_out} + F_{SI} + F_{CS}\right) - F_{SGWD} \tag{27a}$$

$$F_{SGWD} = \left(D_n + D_w + D_t\right)C_{BA} \tag{27b}$$

$$F_{CS} = \left(D_n + D_w + D_t\right)C_{CS} \tag{27c}$$

$$F_{SI} = S\frac{dV_{SI}}{dt} \tag{27d}$$

where V_{BA} is the volume of BA and C_{BA} is the chemical concentration in BA. C_{CS} is the chemical concentration in the ocean and, for the contaminants considered, is usually small compared with C_{BA} and can be neglected. F_{SI} results from seawater intrusion. The chemical adsorbed on sand particles tends to desorb in seawater. Thus, seawater intrusion produces an input flux to BA, and the magnitude of this flux is related to the speed of seawater intrusion and the amount of adsorption S. V_{SI} is the volume of intruded seawater.

Chemicals such as phosphate and ammonia are, in most cases, land-derived pollutants as a result of nutrient leaching from the agricultural fertilizer. Sediments in the freshwater aquifer, as a temporary storage for these chemicals due to high adsorption, become the immediate source of chemicals to the brackish aquifer when seawater intrusion occurs and the chemicals desorb into the brackish groundwater from the sediment. Here, a simulation is presented to illustrate how the local groundwater circulation and oscillations affect the transfer of land-derived pollutants. In the simulation, the FA was assumed to be in an equilibrium state initially and seawater intrusion occurred between $t = 0$ and 10 d. The saltwater front retreated shoreward between $t = 10$ d and 20 d. Other assumptions were: $K_d = 400$. $D_n = 3.75$ m^3/d/m, $D_w + D_t = 90$ m^3/d/m, $dV_s/dt = 5\%D_n$, $C_{in} = 1$ kg/m^3, and $V_{BA} = 500$ m^3/m. During seawater intrusion, the output mass flux from the FA is described by Eq. (26a) and during the retreat of the salt wedge, F_{FA_out} is given by Eq. (26b). The time that it takes for the FA to reach the equilibrium state after the retreat of the salt wedge can be estimated by $F_{SI}t_{SI}/F_{FA\_out}$.

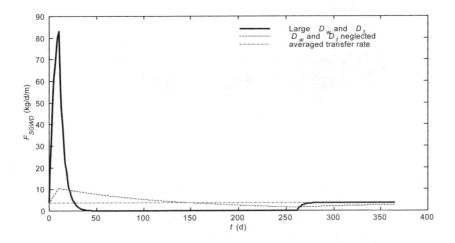

Figure 7: Simulated rates of the transfers of land-derived chemicals to the ocean.

The simulated rate of chemical transfer to the ocean is shown in Figure 7, with the results from a comparison simulation with D_n and D_t neglected. A large increase of the transfer rate is clearly evident as a result of the seawater intrusion and the local groundwater circulation/oscillating flows. The first factor (i.e., seawater intrusion) contributes to an extra and excessive source of the chemical. The second factor (i.e., the local groundwater circulation and oscillating flows) provides the mechanism for rapid flushing of the BA, resulting in increased chemical transfer to the ocean. Without the second factor, the large impulse of chemical input to the ocean would not occur as demonstrated by the comparison simulation (dashed curve in Figure 7). The increase of F_{SGWD} is substantial, more than 20 times as high as the averaged rate. As the salt wedge retreats, the transfer rate decreases to zero since the inland chemical is all adsorbed in the FA. The local processes do not change the total amount of the chemical input to the ocean, which is determined by the inland source.

The tide-induced flushing effect is further illustrated by the following simulation based on a one-dimensional mass transport model,

$$\frac{\partial c}{\partial t} = D_c \frac{\partial^2 c}{\partial x^2} - V \frac{\partial c}{\partial x} \qquad (28a)$$

with

$$V = Ki_n + \sqrt{2} K \kappa A_0 \exp(-\kappa x) \cos\left(\omega t - \kappa x + \frac{\pi}{4}\right) \qquad (28b)$$

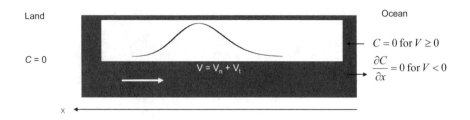

Figure 8: Tidal effects on transport of a contaminant plume.

The first term of the RHS of Eq. (28b) is the net groundwater flow rate and the second term represents the oscillating flow induced by tides (based on the analytical solution, Eq. (2)). The initial concentration is specified according to an existing plume shown in Figure 8. The boundary conditions for the chemical transport are: $c = 0$ at the inland boundary, and $c = 0$ for $V > 0$ and $\partial c/\partial x = 0$ for $V < 0$ at the seaward boundary. The following parameter values are used in the simulation: i_n (regional hydraulic gradient) $= 0.01$, $A_0 = 0$, 1 and 2 m, T (tidal period) $= 0.5$ d, $K = 20$ m/d, $\overline{H} = 10$ m, $n_e = 0.2$, $\alpha = 3$ m ($D_c = \alpha V$), L (distance of the landward boundary from the shore) $= 150$ m.

The results displayed in Figure 9 show that the residence time of the chemical in the aquifer decreases due to tidal oscillations (left-hand side panel of Figure 9). The tidal effects also lead to dilution of the exit chemical concentration significantly (right-hand side panel of Figure 9). Such dilution may reduce the impact of chemicals on the beach habitats.

3.2 Tide-Induced Mixing of Fresh Groundwater and Seawater

In this section, we address how the fresh groundwater discharges to the ocean. Previous studies, neglecting the tidal effects, predict that the freshwater overlies the intruded seawater and discharges to the ocean with little mixing with the saltwater. The limited mixing, driven by the density effects, occurs along the saltwater wedge. A simulation was conducted using SeaWat (http://water.usgs.gov/ogw/seawat/) to examine the tidal effects on the freshwater discharge. Density-dependent groundwater flow in a coastal aquifer subject to tidal oscillations was simulated with a set of parameter values representing the shallow aquifer conditions.

The simulation was run first without the tidal oscillations until a steady state was reached. The result of the salinity distribution in the aquifer shows the traditional view of the groundwater discharge as discussed above (top panel of Figure 10). The tide was then introduced into the simulation, which continued to run for 100 tidal cycles and reached a quasi-steady state. The result shows a very different salinity distribution from that without the tidal effects (middle panel of Figure 10). First, a saline plume was formed in

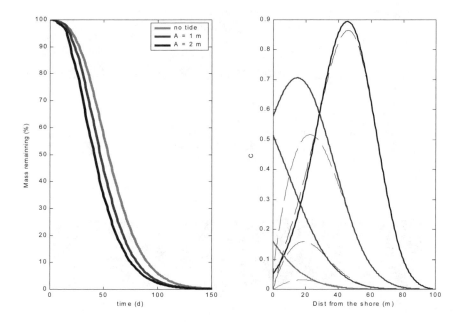

Figure 9: Tidal effects on chemical transport in a coastal aquifer. LHS panel: mass remaining in the aquifer (in percentage of the initial mass) versus time under different tidal conditions. RHS panel: chemical concentration profiles in the aquifer at different times (from the top to the bottom: the results shortly after the simulation started, a bit later, much later, and near the end of the simulation). Dashed lines are for results with tidal effects ($A = 2$ m) and solid lines for results without tidal effects.

the upper part of the beach. The freshwater discharged to the sea through a "tube" between this upper saline plume and the intruding saltwater wedge. Secondly, the freshwater discharge tube contracted and expanded as the tide rose and fell (shown in the attached animation), suggesting considerable mixing activities. Such mixing is also indicated by the salinity gradient shown in the bottom panel of Figure 10. These simulated salinity profiles are consistent with recent results from laboratory experiments [Boufadel, 2000].

    In analyzing the simulated flow and mass transport process, we are particularly interested in (a) how the mean (advective) transport of salinity is affected by the oceanic oscillations, and (b) whether the oceanic oscillations (water exchange) cause diffusive/dispersive transport of salinity. This diffusive transport represents the local, small-scale mixing. For this purpose, the following decomposition is taken,

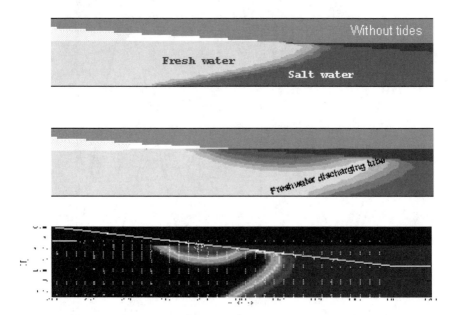

Figure 10: Salinity distribution in the near-shore area of the aquifer. Top panel: without tidal effects; fresh groundwater discharging to the sea without much mixing with underlying seawater. Middle panel: tidal effects leading to the formation of the upper saline plume and the freshwater discharging tube, and considerable mixing between the freshwater and seawater. Color plots are available on the CD.

$$u(x,z,t) = U(x,z) + u'(x,z,t) \qquad (29a)$$

$$c(x,z,t) = C(x,z) + c'(x,z,t) \qquad (29b)$$

where u and c are the raw data of instantaneous flow velocity and salinity; U and C are averaged flow velocity and salinity over the tidal cycle (24 hrs); and u' and c' are the tidally fluctuating flow velocity and salinity. The total mass transport of salinity can then be determined,

$$M = \overline{uc} = UC + \overline{u'c'} \qquad (30)$$

The first term represents the transport due to the mean flow (advection). The second transport component is the diffusive/dispersive flux. In Figure 11, we show calculated mean transport flux and diffusive flux. It is interesting to note that the two fluxes exhibit quite different patterns, and the magnitude of

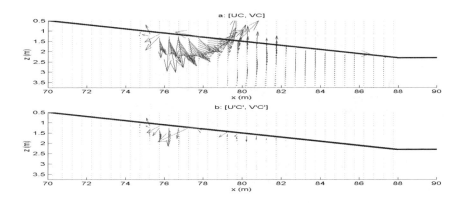

Figure 11: The mean (a) and diffusive (b) mass transport fluxes.

the mean transport flux is one order of magnitude larger than the diffusive flux. Based on the calculated diffusive/dispersive flux, one can estimate the apparent diffusion/dispersion coefficient (the local mixing intensity parameter),

$$D = -\frac{\overline{u'c'}}{\nabla C} \tag{31}$$

where ∇C is the mean salinity gradient.

These results show that the tides affect the near-shore groundwater flow and transport processes significantly, leading to increased exchange and mixing between the aquifer and the ocean. Such effects can alter the geochemical conditions (redox state) in the aquifer and modify the chemical reactions. As shown numerically below, the exchange enhances the mixing of oxygen-rich seawater and groundwater, and creates an active zone for aerobic bacterial populations in the near-shore aquifer. This zone leads to a considerable reduction in breakthrough concentrations of aerobic biodegradable contaminants at the aquifer–ocean interface.

3.3   Tidal Effects on Chemical Reactions

MODFLOW and PHT3D were used to model contaminant transport and biodegradation in coastal aquifers affected by tidal oscillations. Two mobile chemicals were included in the simulation: oxygen as the electron acceptor and toluene as a representative biodegradable contaminant. An aerobic bacterium was included as an immobile phase. The biodegradation process was oxygen-limited (i.e., sufficient substrate). The inland contaminant source was specified at the cells near the water table (Figure 12).

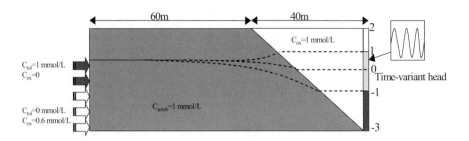

Figure 12: Schematic diagram of the model set-up and boundary conditions implemented in the simulation.

The simulation was run first without tides until a steady state of chemical concentrations was reached. Tides started after that. An animation of the simulation results is contained in the accompanying CD. The image plots of steady state concentrations for toluene, oxygen, and bacteria are shown in Figure 13. Due to the lack of oxygen, little degradation of toluene occurred except in the smearing diffusive layers. Correspondingly, little growth of bacteria can be observed. The chemical concentrations at a high tide after five tidal cycles were shown in Figure 14. The tidal effect is clearly evident: first it created an oxygen-rich zone near the shoreline, which led to biodegradation of toluene. Secondly, it enhanced the mixing process. The smearing layer was thickened. The results at the low tide show similar patterns and changes in the chemical concentrations. In short, the simulation demonstrates that tidal oscillations lead to the formation of an oxygen-rich zone in the near-shore aquifer area. Aerobic bacterial activity sustained by the high O_2 concentration in this active zone degrades the contaminants. These effects, largely ignored in previous studies, may have significant implications for the beach environment.

4. CONCLUSIONS

Coastal water pollution is a serious environmental problem around the world. Most contaminants are believed to be sourced from the land. To develop sound strategies for coastal water pollution control, we must be able to quantify the sources, pathways, and fluxes of contaminants to the coastal zone. Traditionally, terrestrial fluxes of chemicals to coastal water have been estimated on the basis of river flow alone. Recent studies suggest that contaminants entering the coastal zone with groundwater discharge can significantly contribute to coastal marine/estuarine pollution.

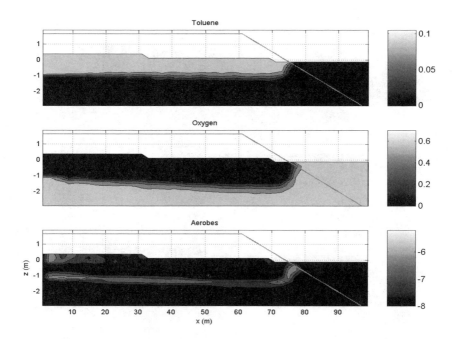

Figure 13: Image plots of the steady state concentrations for toluene (top panel), oxygen (middle panel), and bacteria (bottom panel). Color plots are available on the CD.

To determine the fluxes of chemicals to coastal water, it is important to quantify both the chemical transport processes and reactions on the pathway. There has been a large amount of research work conducted on how the chemicals may be transformed during the transport along the surface pathway, i.e., the role of a surface estuary. In contrast, little is known about the chemical transformation in the near-shore area of a coastal aquifer prior to chemicals' discharge to coastal water.

In this chapter, we first reviewed a large volume of work on tide-induced groundwater oscillations in coastal aquifers, focusing on analytical solutions of the tidal water table fluctuations. In the second part, we discussed the effects of tides and other oceanic oscillations on the chemical transport and transformation in the aquifer near the shore, drawing an analogy to the surface estuary–subsurface estuary. The discussion, based on several on-going studies, illustrated the important role that a subsurface estuary may play in determining the subsurface chemical fluxes to coastal waters. Although the tidal influence on the water table dynamics has been

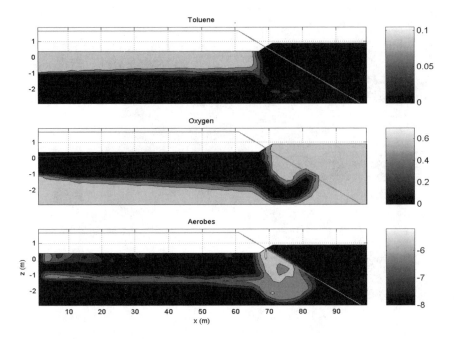

Figure 14: Image plots of the concentrations for toluene (top panel), oxygen (middle panel) and bacteria (bottom panel) at the high tide after five tidal cycles. Color plots are available on the CD.

subjected to numerous studies, the effects of tides on the fate of chemicals in the aquifer have not been investigated adequately. Quantification of these effects is clearly needed in order to

- provide better understanding of the pathway of land-derived nutrients and contaminants entering coastal waters; and
- provide useful information for integrating the management of upland and lowland catchments, and for improving strategies for sustainable coastal resources management and development.

Acknowledgments

Research work carried out by the authors on chemical transport and transformation in coastal aquifers was supported by the Leverhulme Trust (UK) under project F/00158/J.

REFERENCES

Ataie-Ashtiani, B., R.E. Volker, and D.A. Lockington, "Tidal effects on sea water intrusion in unconfined aquifers," *J. Hydrol.*, **216**, 17–31, 1999.

Baird, A.J., T. Mason, and D.P. Horn, "Validation of a Boussinesq model of beach groundwater behaviour," *Mar. Geol.*, **148**, 55–69, 1998.

Barry, D.A., S.J. Barry, and J.-Y. Parlange, "Capillarity correction to periodic solutions of the shallow flow approximation," In *Mixing Processes in Estuaries and Coastal Seas*, C. B. Pattiaratchi (ed.), AGU, Washington, DC, 496–510, 1996.

Bear, J., *Dynamics of Fluids in Porous Media*, Elsevier, New York, 1972.

Bokuniewicz, H., "Groundwater seepage into Great South Bay, New York," *Estuar. Coastal Mar. Sci.*, **10**, 437–444, 1980.

Boufadel, M.C., "A mechanistic study of nonlinear solute transport in a groundwater-surface water system under steady and transient hydraulic conditions," *Water Resour. Res.*, **36**, 2549–2565, 2000.

Buddemeier, R.W. (Ed.), *Groundwater discharge in the coastal zone*, LOICZ/R&S/96-8, Texel, The Netherlands, 1996.

Burnett, W.C., M. Taniguchi, and J. Oberdorfer, "Measurement and significance of the direct discharge of groundwater into the coastal zone," *J. Sea Res.*, 106–116, 2001.

Cartwright, N., and P. Nielsen, "Groundwater dynamics and salinity in coastal barriers," *Proc. 1st Int. Conf. Saltwater Intrusion &Coastal Aquifers*, D. Ouazar and A.H.-D. Cheng (eds.), April 23–25, Essaouira, Morocco, 2001.

Cooper, H.H., Jr., "A hypothesis concerning the dynamic balance of fresh water and salt water in a coastal aquifer," *J. Geophys. Res.*, **71**, 4785–4790, 1959.

Dracos T., Ebene nichtstationare Grundwasserabflusse mit freier Oberflache. Swiss Federal Technical Laboratory of Hydraulic Research and Soil Mechanics, *Rep. No. 57*, p. 114., 1963.

Enot, P., L. Li, H. Prommer, and D.A. Barry, "Effects of oceanic oscillations on aerobic biodegradation in coastal aquifers," *Geo. Res. Abs.*, European Geophysical Society, 3: 18, 2001.

Haynes, D., and K. Michael-Wagner, "Water quality in the Great Barrier Reef world heritage area: Past perspectives, current issues and new research directions," *Mar. Pollut. Bull.*, **41**, 7–12, 2000.

Huyakorn, P.S., P.F. Andersen, J.W. Mercer, and H.O. White, "Salt intrusion in aquifers: Development and testing of a three-dimensional finite element model," *Water Resour. Res.*, **23**, 293–319, 1987.

Jeng, D.S., L. Li, and D.A. Barry, "Analytical solution for tidal propagation in a coupled semi-confined/phreatic coastal aquifer," *Adv. Water Resour.*, **25**(5): 577–584, 2002.

Jiao, J.J., and Z.H. Tang, "An analytical solution of groundwater response to tidal fluctuation in a leaky confined aquifer," *Water Resour. Res.*, **35**, 747–751, 1999.

Johannes, R.E., "The ecological significance of the submarine discharge of groundwater," *Mar. Ecol. Prog. Ser.*, **3**, 365–373, 1980.

Li, H.L., and J.J. Jiao, "Analytical solutions of tidal groundwater flow in coastal two-aquifer system," *Adv. Water Resour.*, **25**, 417–426, 2002a.

Li, H.L., and J.J. Jiao, "Tidal groundwater level fluctuations in L-shaped leaky coastal aquifer system," *J. Hydrol.*, **268**, 234–243, 2002b.

Li, L., D.A. Barry, and C.B. Pattiarachi, "Numerical modelling of tide-induced beach water table fluctuations," *Coastal Eng.*, **30**, 105–123, 1997a.

Li, L., D.A. Barry, J.-Y. Parlange, and C.B. Pattiaratchi, "Beach water table fluctuations due to wave runup: Capillarity effects," *Water Resour. Res.*, **33**, 935–945, 1997b.

Li, L., D.A. Barry, F. Stagnitti, and J.-Y. Parlange, "Submarine groundwater discharge and associated chemical input to a coastal sea," *Water Resour. Res.*, **35**, 3253–3259, 1999.

Li, L., and D.A. Barry, "Wave-induced beach groundwater flow," *Adv. Water Resour.*, **23**, 325–337, 2000.

Li, L., D.A. Barry, F. Stagnitti, J.-Y. Parlange, and D.S. Jeng, "Beach water table fluctuations due to spring-neap tides," *Adv. Water Resour.*, **23**, 817–824, 2000.

Li, L., P. Enot, H. Prommer, F. Stagnitti, and D.A. Barry, "Effects of near-shore groundwater circulation on aerobic biodegradation in coastal unconfined aquifers," *Proc. 1st Int. Conf. Saltwater Intrusion & Coastal Aquifers*, D. Ouazar and A.H.-D. Cheng (eds.), April 23–25, Essaouira, Morocco, 2001.

Moore, W.S., "Large groundwater inputs to coastal waters revealed by ^{226}Ra enrichment," *Nature*, **380**, 612–614, 1996.

Moore, W.S., "The subterranean estuary: a reaction zone of ground water and sea water," *Mar. Chem.*, **65**, 111–125, 1999.

Nielsen, P., "Tidal dynamics of the water table in beaches," *Water Resour. Res.*, **26**, 2127–2134, 1990.

Nielsen, P., R. Aseervatham, J.D. Fenton, and P. Perrochet, "Groundwater waves in aquifers of intermediate depths," *Adv. Water Resour.*, **20**, 37–43, 1997.

Nielsen, P., "Groundwater dynamics and salinity in coastal barriers," *J. Coastal Res.*, **15**, 732–740, 1999.

Parlange J.-Y., F. Stagnitti, J.L. Starr, and R.D. Braddock, "Free-surface flow in porous media and periodic solution of the shallow-flow approximation," *J. Hydrol.*, **70**, 251–263, 1984.

Parlange, J.-Y., and W. Brutsaert, "A capillary correction for free surface flow of groundwater," *Water Resour. Res.*, **23**, 805–808, 1987.

Raubenheimer, B., R.T. Guza, and S. Elgar, "Tidal water table fluctuations in sandy beaches," *Water Resour. Res.*, **35**, 2313–2320, 1999.

Simmons, G.M., "Importance of submarine groundwater discharge and seawater cycling to material flux across sediment/water interfaces in marine environments," *Mar. Ecol. Prog. Ser.*, **84**, 173–184, 1992.

Turner, I., "Water table outcropping on macro-tidal beaches: A simulation model," *Mar. Geol.*, **115**, 227–238, 1993.

Turner, I.L, B.P. Coates, and R.I. Acworth, "Tides, waves, and super-elevation of groundwater at the coast," *J. Coastal Res.*, **13**, 46–60, 1997.

Wang, J. and T.-K. Tsay, "Tidal effects on groundwater motions," *Trans. Porous Media*, **43**, 159–178, 2001.

Younger, P.L., "Submarine groundwater discharge," *Nature*, **382**, 121–122, 1996.

Zekster, I.S., and H.A. Loaiciga, "Groundwater fluxes in the global hydrological cycle: past, present and future," *J. Hydrol.*, **144**, 405–427, 1993.

Zhang, Q., R. E. Volker, and D. A. Lockington, Influence of seaward boundary condition on contaminant transport in unconfined coastal aquifers, *J. Contam. Hydrol.*, **49**, 201–215, 2001.

CHAPTER 7

Determination of the Temporal and Spatial Distribution of Beach Face Seepage

D.W. Urish

1. INTRODUCTION

Man is a creature closely linked to the coastal areas for many reasons. Some 70% of the earth's population live within coastal zones, with the large portion of that population within a few kilometers of saltwater. Historically, as well as today, the saltwater seas are the main access to both the products of seas, as well as the lands beyond, a natural location for the development of commerce, habitation, and industrialization. This heavy concentration of mankind and his activities creates many anthropogenic products detrimental to the environment and to man himself. Much of this environmental impact moves into the groundwater system as a natural consequence of the hydrologic cycle. The impact of civilization is most keenly recognized in the more confined and poorly flushed estuaries, bays, and coastal lagoons.

Within the larger concept of global water budgets, all freshwater falling on the terrestrial components of the earth eventually returns to the "mother of waters," the saltwater seas. The path of a molecule of water may be long and tenuous following varying hydraulic gradients until it finally reaches its original source and the hydrologic cycle repeats. The meeting of freshwater with saltwater may be a glacier caving its icebergs into the sea, mighty rivers, or in our area of interest the more subtle, but constant discharge of coastal fresh groundwater. The time of transient through the ground may range from many years for coastal plains and large peninsulas to days for small islands and near-shore recharge. But eventually it reaches the saltwater, carrying with it many terrestrial derived components, both natural and anthropogenic. The increased recognition of the importance of the coastal groundwater discharge zone, and the greatly increased capabilities for

1-56670-605-X/04/$0.00+$1.50
© 2004 by CRC Press LLC

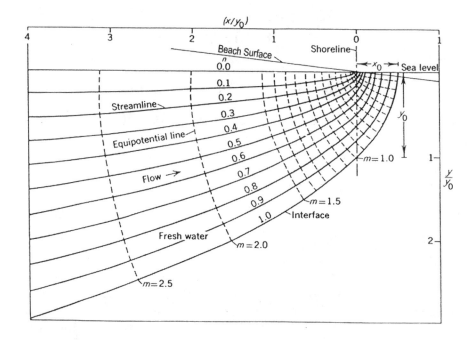

Figure 1: Fresh groundwater flow and discharge pattern
(after Glover [1964]).

data collection and analysis, have encouraged the study of the dynamic
aspects of tidal effects for coastal groundwater seepage analysis [Gilbin and
Gaines, 1990; Millham and Howes, 1994; Portnoy *et al.*, 1998].

The objective of this discussion is to describe the dynamic concept
of the coastal freshwater–saltwater relationship and the techniques that can
be used to determine coastal fresh groundwater seepage in a quantitative and
qualitative form. The descriptions and methods described are primarily
directed to the more quiescent shores of the relatively sheltered bays and
lagoons, and generally the source of most critical environmental concerns. It
is further most applicable to the sandy seashore, influenced by the changing
water levels of the ocean tides. In many cases a sandy beach or cove, even on
the rock bound coast, is the zone of primary fresh groundwater discharge.

2. CONCEPTS

2.1 Freshwater-Saltwater Relationships

Where freshwater meets saltwater in a permeable landmass, the
freshwater will tend to float on the more dense saltwater according to the
Ghyben-Herzberg Principle [Drabbe and Ghyben, 1889; Herzberg, 1901]. In

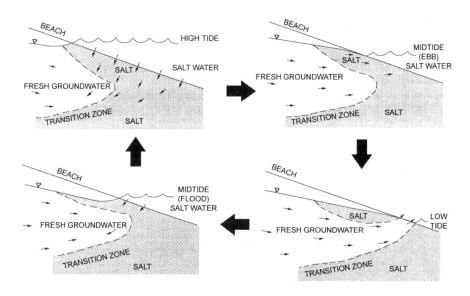

Figure 2: The sequence of coastal groundwater discharge through a sandy
beach during the tidal cycle.

an insular landmass, such as an island or peninsula, this configuration of
body of freshwater will approximate a lens, bounded by and underlain by
saltwater. The coastal manifestation of this lens is a pinching out of the lens
at the coastal boundary to discharge through a narrow zone at the tidal
margin described in a steady state theoretical case by Glover [1964], and as
further illustrated for a coastal margin in Figure 1.

Delineation of coastal discharge is a much more elusive problem
when one considers the changing groundwater conditions in the inter-tidal
zone incorporating the complexities of a boundary which changes cyclically
twice a day both laterally and vertically, highly variable salinity, fluctuating
hydraulic heads, and a geologically heterogeneous beach [Turner, 1993a;
Baird and Horn, 1996; Robinson and Gallagher, 1999; Li et al., 2000].

2.2   The Moving Boundary

In tidally influenced coastlines both the freshwater lens and the
discharge patterns are greatly changed from a static condition, depending on
the topography and geologic nature of the beach inter-tidal zone. The water
table in the coastal groundwater moves up and down with the tide;
concurrently the boundary on a sloping beach surges shoreward and seaward;
the beach is flooded with saltwater twice a day, and in many cases the
hydraulic discharge gradient itself changes direction, an extremely complex
and dynamic situation. The basic process of coastal groundwater discharge

through an idealized homogeneous sandy beach during a tidal cycle is illustrated in Figure 2.

During high tide, groundwater flow is hydraulically blocked, with a reverse hydraulic gradient toward the land imposed by the tide, which is higher than the near-shore water table; additionally, saltwater will infiltrate into the land surface adding to and mixing with the fresh groundwater in the beach.

As the tide ebbs the hydraulic gradient reverses and groundwater flow consisting of both salt and freshwater moves toward the lower beach. As low tide approaches groundwater discharge occurs, both as beach face seepage and lower beach submarine discharge. With the rising tide a reverse hydraulic gradient is again established and the groundwater discharge ceases. The cycle then repeats.

Field sampling of coastal groundwater discharge is greatly complicated by the transient nature of the tidally induced changing boundary. The timing and location of the quality of groundwater in three dimensions becomes critical for groundwater sampling. This is further complicated by the indistinct and changing salinity of the beach groundwater and discharge. The earliest freshwater lens models made no attempt to discretely character the hydraulic and chemical nature of the seepage, treating it as a fixed sharp line in time and space.

A significant advancement was the theoretical formulation of the discharge gap representation to describe coastal seepage by Glover [1959] and further described by Cooper [1965] under steady state conditions. This, however, failed to take into account anything other than the assumed discharge without regard for the salinity of the discharge. The distribution of the discharge as a decreasing exponential pattern was first examined in a field setting on the shores of Long Island by Bokuniewicz [1980, 1992], referencing earlier freshwater lake seepage studies by McBride and Pfannkuch [1975]. These field observations, however, were under essentially tideless conditions.

Because of the laterally moving boundary on a sloping beach, there is a much wider outflow gap as well as major changes in the flow pattern of the discharge, including in many cases a complete reversal of flow and salinity. A beach face model, SEEP, was developed by Turner [1993a] to analyze and predict the exit dynamics of groundwater seepage with a falling tide. Turner further describes the role of the capillary fringe in the total water content of the beach.

2.3   Beach Slope Effect

While the determination of mean sea level (MSL) in the open coastal water system is a necessary base line, it should be recognized that in a sloping beach there is a dynamic phenomenon caused by the tide movement which can create an "effective mean sea level" (EMSL) in the beach considerably above open water measured MSL [Urish and Ozbilgin, 1989]. This was later elaborated on by Nielsen [1990] and Hegge and Masselink [1991]. The seawater is mounded in the upper beach by the dynamic movement of tide and consequent infiltration of saltwater as it moves up the beach face. There is, in effect, a pumping action caused by rapid infiltration of the seawater in the upper beach during high tide and much slower drainage of the seawater through the lower beach at low tide. This results in a super elevation of the apparent sea level boundary condition, which has been measured as much as 0.5 feet above open water MSL for a 5 foot tide range on a 0.05 beach slope [Urish, 1980]. This becomes important in modeling coastal boundary conditions.

The inter-tidal beach is subjected to seawater flooding and infiltration from the rising tide, which is then a substantial component of the beach discharge. The rising edge of the incoming tide advances shoreward faster than the discharging freshwater can rise. Thus, the seawater quickly fills the available pore space in the sands of the upper beach, sometimes rising rapidly enough to trap air under the surface. The quantity of infiltrated saltwater in the beach which becomes seepage depends on the residual water content from the previous saturation episode, as well as the downward directed hydraulic gradient. The residual water in the upper portion of the inter-tidal zone is usually a layered mixture of saltwater over freshwater with some mixing, depending on the magnitude of the freshwater discharge and the antecedent drainage characteristics of the beach.

As Bokuniewicz [1992] points out, however, saline pore water overlying fresh pore water has an inherently unstable density gradient, causing "fingering" of the different densities of water to occur; this leads to greater uncertainty in any attempts at determining the volume of infiltrated saltwater directly. The presence, however, of a substantial layer of infiltrated saltwater overlying freshwater in the inter-tidal zone is well established by both direct water table sampling [Portnoy et al., 1998] and by indirect surface electrical resistivity soundings in the inter-tidal beach [Frohlich, 2001].

3. METHODOLOGY

3.1 Elevation Measurements

3.1.1 Elevation Control and Datums

In order to relate water levels to the beach and near-shore surfaces it is essential that beach topographic profiles be made and referenced to a fixed datum, the same as used for setting elevation reference points on monitor wells and tidal stations in the study area. While more sophisticated (and expensive) survey methods such as the "total station" may be used, for the limited area usually involved, the "automatic level" and tape are generally most efficient. The most frequently used reference datum is the 1929 National Geodetic Vertical Datum (NGVD29) or more recent North American Vertical Datum of 1988 (NAVD88), which can be related for a specific geographic area to the NGVD29 by an adjustment constant. While the NGVD29 datum is frequently referred to as "mean sea level", it is only a very crude approximation, and is far from the precision necessary for coastal water level measurements.

Complicating coastal elevation measurements is the fact that tide table predictions and tide station measurements are usually reference to locally determined assigned datums of mean lower low water (MLLW). This is a datum determined as zero from the average of the lower of the two low waters of each day for the past 19 years. For the United States the values in popular references such as Reed's Nautical Almanacs [Herzog, 2003] are still in feet, rather than the more globally accepted meters. Tide level predictions for specific locations can also be obtained from the National Oceanic and Atmospheric Administration (NOAA) web site www.co-ops.nos.noaa.gov [Wolf and Ghilani, 2002]. The correction necessary to convert the local MLLW value to a 1929 NGVD or 1988 datum can be obtained from the web site. For example, for the Narragansett Bay 2.92 feet must be subtracted from the MLLW value of tide to obtain the equivalent water level relative to the NGVD 1929 Datum. This is necessary information for coastal field investigation planning and coastal engineering.

3.1.2   Water Level Measurements

Water level measurements taken to a precision of 0.03 m and referenced to a datum are essential to any study of groundwater in order to evaluate the transport and movement characteristics of the groundwater, the receiving water, and tidal systems. These water level measurements are generally used as direct measurements of hydraulic heads and piezometric pressures.

Any number of water level measurement (depth to water) techniques can be used depending on the length of time of the investigation, the precision required and the resources available. The following discussion is divided into the general categories of short term and long term. It is intended to be comprehensive, but is most specifically not inclusive of all possible techniques.

In all cases it is important to recognize the importance of concurrently determining the density of the water in the monitor wells being measured, usually determined indirectly as a function of measured salinity. The concept of variable water density as it relates to groundwater flow systems is explained in excellent detail by Lusczynski [1961]. All water level measurements must be converted to freshwater or saltwater equivalents in order to evaluate the water levels as hydraulic heads. To make this conversion both the depth of the water column in the monitor well as well as the salinity (density) must be known. As an example, the measured water level in a monitor well with a column of 3.00 m of saltwater with a density of 1.020 must be increased by 0.06 m to be a freshwater equivalent for comparison with the heads in freshwater monitor wells.

3.1.2.1 Short-term water level measurements

Manual point-in-time depth to water measurements can be accomplished in monitor well or tidal stilling wells by several methods. Once the depth to water is determined from the top of a well casing with known elevation referenced to a datum, subtraction of that value from the well casing elevation gives the water level elevation. This can be done by direct water level measurement with a tape in shallow wells and by the "wetted tape" method or with electrical response devices in deeper wells.

3.1.2.2 Long-term water level measurements

In many cases the field study requires a long-term continuous series of measurements, which may extend into months. In other cases it is necessary to collect data from many wells simultaneously and at very short intervals of time. For such cases it is not feasible, if not impossible, to collect data points manually. For this purpose mechanized or computerized data collection is necessary:

A) The oldest method is the drum water level recorder in which a float is connected mechanically to a time oriented rotating drum. A pen in the recorder then traces the track of water fluctuation on graph paper placed on the rotating drum. As might be expected there are many opportunities for recorder failure; among other things, the pen may run out of ink, the power source may run out, the float may get fouled, etc. The benefit is that with well-maintained equipment and frequent performance checks it gives a direct visual plot of results. The graphic plot then must be manually converted to digital values for further analysis.

B) The most commonly used method is the hydraulic pressure transducer. This consists of a computer data logger connected by cable to a small diameter pressure transducer probe that can be placed in the well. The probe measures the water level by the pressure changes on a very small diaphragm that then transmits electrical signals of its movement to the data logger. The water level is actually measured as the weight of water above a carefully elevation-referenced transducer. It is apparent then that the calibration of the transducer must be corrected for the density of water; e.g., if a transducer calibrated for freshwater use is placed under 4.00 m of sea water at a density of 1.025, rather than freshwater at a density of 1.000, then the logger reading will be 4.10 m rather than 4.00 m, a very significant difference in groundwater measurements. The logger unit can be programmed for timing and frequency of data collection and downloaded directly into a computer file.

C)   A more recent automatic water level recorder especially suitable for shallow systems is the "Ecotone" capacitance water level monitoring instrument, manufactured by Remote Data Systems, which uses an electrical wire capacitor method. This requires a special tube or monitor well and so is not as adaptable as the pressure transducer, which can be placed in any well, but does have the advantage that each well is a self-contained unit and so can be placed in widely separated remote locations. Further, it is not affected by water density and barometric pressure. As with the pressure transducer logger, it can be programmed for frequency interval of data collection and downloaded directly into a computer file.

3.2   Beach Sediments and Topography

Recognizing that the hydraulic conductivity of beach sediments may vary greatly both horizontally and vertically, it is very useful to take soil samples at different locations along the beach to characterize the beach and its variability. Undisturbed samples should be collected during low tide in tubes pressed into the walls and bottoms of excavations to obtain both horizontal and vertical oriented samples. If a disturbed sample is all that can be obtained, then care should be taken to compact it to a maximum density to approximate the in-situ condition before running permeability tests. In this case it should be recognized that the inherent in-situ anisotropy, which may range from 5 to 50 for beach samples, is lost in the reconstituted sample. It is possible to assume a value for anisotropy and back calculate probable values for K_h and K_v using the relationship $K = (K_h K_v)^{1/2}$. If a reasonable value of 10 is taken, then $K_h = 3.16\,K$ and $K_v = K/3.16$.

It is necessary to determine the beach profile to understand the relationship of measured water table and tide levels to the beach surface through which seepage occurs. The profile should be referenced with horizontal and vertical control in order that subsequent beach surveys can be related to the same fixed reference. Beach surfaces are far from stable, changing with each tidal cycle and more dramatically with storms. For long-term studies a number of profiles need to be accomplished.

3.3   Coastal Seepage Measurements

3.3.1   Thermal Infrared Aerial Imagery

Thermal infrared imagery has been a particularly useful tool to determine coastal fresh groundwater discharge patterns and specific locations. The proper application of the technique, however, requires careful attention to the timing of coastal groundwater discharge. In a beach composed of permeable porous media the timing of the imaging survey must occur during the period ½ hour before to 1 hour after low tide, during the period of maximum fresh groundwater discharge. It should be noted, however, that there are some hydrogeologic exceptions to this general rule, namely in coastal environments where a beach confining or semi-confining layer may preclude open phreatic discharge through the beach. In such a case the water table will be elevated by a rising tide and discharge may take place at high tide in the upper beach at the upper limit of the confining layer; only a detailed on-site survey can ascertain if such a hydrogeologic condition exists in the areas of interest.

The thermal infrared method maps temperatures of surfaces exposed to a super-cooled detector, which is mounted on a small aircraft. The results can be visually interpreted to identify groundwater discharge along a coastal margin by measuring the difference in thermal spectral response of the water along the coast. The temperature contrast can be either a colder groundwater to warmer receiving water as occurs in the late summer or warmer groundwater to colder receiving water as occurs in the winter months. For a successful thermal imagery survey the groundwater-receiving water temperature contrast should be no less than about 5°C. The ability to detect the colder groundwater is further enhanced by the tendency of the less dense freshwater to float on the top of saltwater. In a summer survey the colder fresh groundwater appears as a dark plume emanating from the shore. There should be two flight runs accomplished approximately ½ to 1 hour apart in order to distinguish between fixed coastal features, which also may give a thermal response, and the moving plumes of discharging freshwater. This is illustrated in Figures 3 and 4, which show images of a moving freshwater plume taken one hour apart during low tide.

3.3.2   Beach Salinity Transects

Beach salinity sampling transects can be made transverse and parallel to the beach line at low tide to ascertain the variability of quality of seepage in a local zone. Such sampling must be at closely spaced locations, but because the quality and location of the water changes with time it is necessary that the sampling be done very rapidly. This is best done by

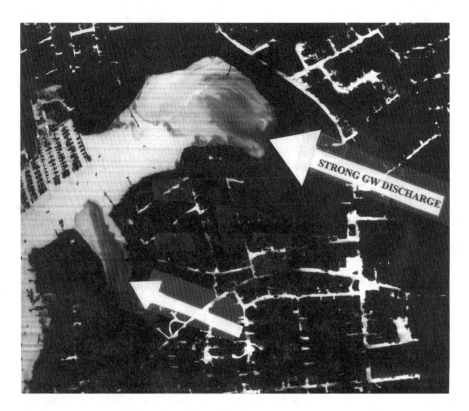

Figure 3: Thermal infrared image of fresh groundwater plume (image one).

extracting small water samples at shallow depths with a small probe attached to a manually operated syringe. The small quantity thus obtained can then be rapidly analyzed for salinity using a small handheld refractometer.

3.3.3   Direct Beach and Coastline Water Quality Sampling

The selection of a proper method for groundwater sampling in the beach environment depends on the intent and duration of the survey, and implicitly the available resources. It is important to recognize that all direct sampling methods are point measurements and hence may not be representative of seepage over a broader regional shoreline because of the great heterogeneity of the coastal discharge zone. Field measurements of piezometric heads as well as low tide beach observations indicate that a substantial amount of discharge occurs under both subaerial and submarine conditions. An additional consideration is that single point sampling may be completely out of a primary freshwater seepage zone even though substantial discharge may occur. Thus one should consider a broader based

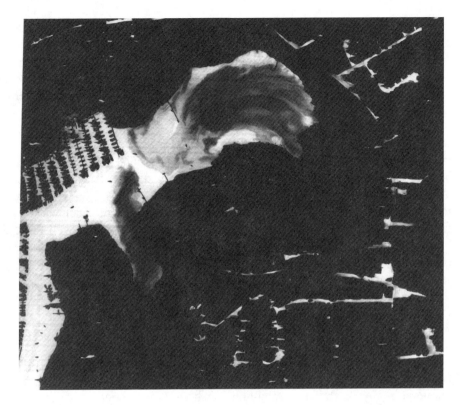

Figure 4: Thermal infrared image of freshwater plume (image two), 1 hour after that of Figure 3.

reconnaissance such as the thermal infrared imagery, or at least rapid shoreline transects, to identify zones of probable fresh groundwater discharge before detailed sampling is undertaken.

3.3.3.1 Short-term sampling

Short-term sampling to characterize the nature of coastal groundwater in three dimensions can be done by direct sampling with probes and by seepage meters. Discrete groundwater sampling can be done both rapidly by shallow probes going only a few centimeters into the seepage face, by deeper hand-driven probes going as deep as 5 m, or even deeper by power procedures.

In the submarine part of the discharge zone seepage meters can be used. Seepage meters are limited to sampling for submarine seepage since they must remain under water. While more sophisticated electronic devices are now coming on the market, most seepage meters have two basic components, namely, a shallow pan usually no larger than a meter which is inverted over the area to be sampled, and a seepage bag placed on a stopcock set in the inverted pan [Lee, 1977]. The seepage water then flows through the confined space of the inverted pan and accumulates in the bag. The amount of water collected over a determined period of time can then be measured and seepage rate calculated. More recent innovations of the seepage meter have been made to accomplish automated continuous flow measurements

3.3.3.2 Long-term sampling

Long-term sampling at fixed locations is best done utilizing properly installed monitor wells for both water quality and water level measurements. While good monitor wells can be installed by hand methods, it is frequently more expedient to contract a well driller, preferably from a geotechnical firm familiar with the purpose and technical specifications for monitor wells. The best drilling method employs the hollow stem auger which permits the obtaining of relatively undisturbed split spoon samples as well as water samples at specific depths during the drilling process. In order to ascertain the vertical distribution of water quality and piezometric heads a nest of at least three monitor wells needs to be installed at each location.

3.3.4   Water Quality Measurements

The primary parameters of interest in field measurement to locate coastal fresh groundwater seepage are electrical conductivity, salinity, and temperature. After the best sampling locations are ascertained, additional conventional field measurements such as pH and oxygen can be taken, and samples collected, preserved, and conveyed to a laboratory for chemical analysis to any degree of sophistication desired.

The best all around instrument for the exploration phase of seepage investigation is the YSI temperature-conductivity-salinity meter. This is rugged and versatile, and while limited in precision relative to fixed laboratory instruments, is quite suitable for ascertaining if seepage water is fresh or salty. The refractometer is a very convenient instrument for rapid measurements of limited precision; this can determine salinity only to ppt, but can provide a reading very rapidly and requires only a drop of water.

4.   CASE STUDY [URISH AND QANBAR, 1997]

4.1   Study Location

The study was conducted along the beaches of the Nauset Marsh embayment (Figure 5), a 945 ha back-barrier estuary on Cape Cod, MA connected by an inlet to the Atlantic Ocean.

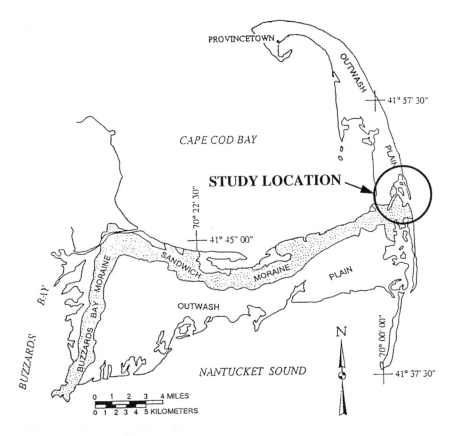

Figure 5: Study location for coastal fresh groundwater seepage.

The surficial deposits are largely unconsolidated glacial sediments deposited by glacial ice and melt water at the close of the Wisconsin Glaciation, some 10,000 years ago. Very old granitic bedrock lies about 170 m below the surface. The beaches are composed of marine reworked shoreline deposits, predominantly of relatively uniform quartz composition ranging from silt to coarse sands. The Nauset Marsh embayment is dominantly medium to coarse sands, with thin upper layers of silt at some locations. In the beaches investigated there was a median grain size diameter range of 0.40 to 1.00 mm and a D_{10} size ("effective size") of 0.10 to 0.36 mm.

The topography of the study area is undulating with elevations ranging from sea level to 4.25 m. There are numerous bays, coves, and coastal wetlands. Surface streams are infrequent with much of the precipitation infiltrating into the sandy soil. The climate of the region is a

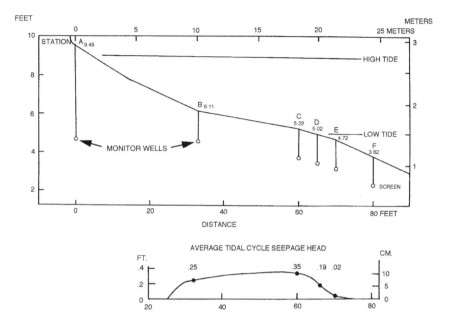

Figure 6: Profile of beach face showing location of monitoring wells and
seepage during a tidal cycle.

maritime humid temperate climate. The average annual rainfall of 110 cm is
evenly distributed throughout the seasons. The aquifers are phreatic with the
groundwater occurring as a freshwater lens "floating" on the denser
underlying salt water. The Nauset Marsh complex is a shallow marine
environment with tides in the 1- to 2-m range, averaging about 1.34 m.
Salinity in the central parts of the water bodies is near that of the connecting
Atlantic Ocean, in the range of 25 to 30 ppt; the near-shore salinities are less,
being strongly influenced by the discharging freshwater, particularly during
low tide periods.

4.2   Methodology

Sets of monitor wells consisting of 3.2 cm inside diameter PVC pipe
with 7.6 cm screen at the lower ends were installed in the beach zone for
piezometric measurements and water quality sampling as shown in Figure 6.
These were placed by hand augering with the center of the screens set 45 cm
below the beach face and below the lowest position of the water table. Water
levels were measured both by direct tape measurements and by pressure
transducers placed in the monitor wells, as well as in the open water for tidal
measurements. Hydraulic head values were corrected for density variation to
freshwater equivalent heads as appropriate [Kohout, 1961; Lusczynski,
1961].

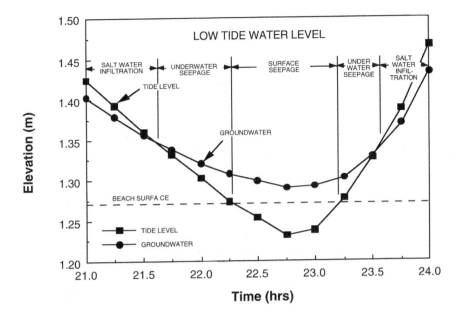

Figure 7: Plot of tide and groundwater levels under low tide conditions.

Low tide shoreline reconnaissance sampling for groundwater discharge salinity was accomplished using a 2 mm internal diameter stainless steel probe with a fine screen tip pushed about 10 cm into the sediments, and the water drawn by vacuum into a syringe set on the tube's upper end. Salinity was determined using a handheld refractometer in the field and by YSI instrument in the lab for collected water samples. Values were standardized to 25°C. Soil samples were also taken in seepage areas and analyzed for grain size using sieve analysis and hydraulic conductivity by the falling head permeameter test, both by ASTM Standards.

The elevations of all monitor wells were established by standard leveling techniques using a TOPCON automatic level and referenced to 1929 National Geodetic Vertical Datum (NGVD), or to an arbitrary local datum where a NGVD benchmark was not available.

4.2.1   Infiltration and Seepage Mechanism

In order to examine the seepage dynamics in detail during a tidal cycle field studies were made at two sandy beach sites in the Nauset Marsh estuary complex. Monitor well water level measurements in the beach for the low tide phase (Figure 7) illustrate the relationships between beach groundwater and tidal water during low and high tide episodes. Groundwater piezometric head measurements at one monitor well at the low tide line and surface water elevations were monitored at 15-min intervals for 7 days using pressure transducers attached to automatic data loggers. Using this information, episodes of high and low tide were selected for detailed seepage analysis.

Results show that for saltwater infiltration to occur two conditions need to exist, namely, 1) The saltwater level in the open water must be higher than the ground surface elevation that it floods, and 2) The saltwater level has to be higher than the groundwater hydraulic head at the monitor well location in the flooded beach. In this case the analysis must examine both subaerial and submarine hydraulic conditions in the beach as the tide recedes past the monitor well location. It is observed that at 21.0 hours infiltration is still occurring, but beginning at 21.6 hours the groundwater head becomes higher than the tide water, but both are higher than the beach surface, thus underwater seepage occurs. At about 22.5 hours the tide level falls below the beach surface at the monitor well, but since the piezometric head of groundwater is greater than the beach surface elevation, surface seepage exists. This continues until 23.2 hours when the beach location is again flooded by a rising tide and submarine seepage occurs. At 23.6 hours seepage ceases and infiltration begins again.

4.2.2   Temporal and Spatial Pattern of Seepage

Using the analytical process described in the preceding section for one monitor well location on the beach, full beach seepage analysis at three sites was accomplished using sets of monitor wells installed transverse to the beach.

A time sequence of plots (Figure 8) showing the magnitude of net hydraulic heads along the beach face over the lower part of the tidal cycle provides insight into the temporal and spatial pattern of seepage as the tide moves down and back up the beach face.

The shaded areas under the curve denote submarine seepage where the tide covers the beach. The sequence starts with all locations showing submarine seepage at 9.0 hours; there was no seepage at 8.0 hours. It is to be noted that the location and magnitude of greatest seepage changes with time, generally moving seaward with the tide. Finally, the sequence ends with all locations showing infiltration.

When the time sequence of seepage is recalculated as average seepage during the tidal cycle, the result of seepage distribution relative to the beach face is as shown in the bottom part of Figure 6. The seepage indicated is composed of both fresh groundwater and infiltrated saltwater.

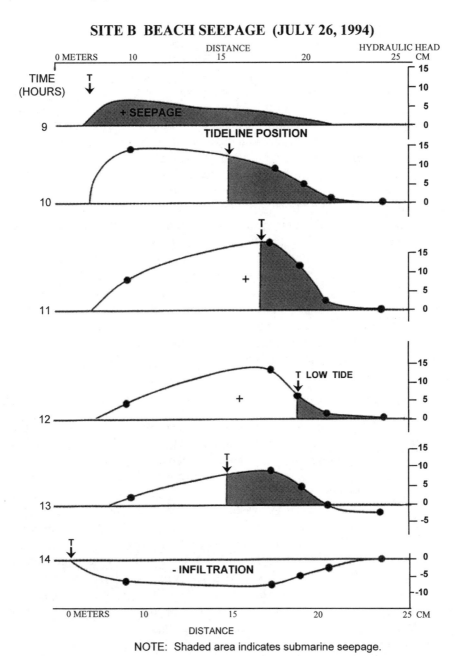

Figure 8: Temporal and spatial sequence of beach seepage and infiltration.

4.2.3   Quality of Seepage

The salinity of the seepage varies with location and time. The freshest measured discharge was 2 ppt, which occurred during low tide at the location of maximum seepage, while the highest salinity of 30 ppt was at the beginning of the discharge period. At all sites the lower part of the seepage zone displayed minimum salinity especially during the lowest part of the tidal cycle. The early discharge includes considerable flushing of saltwater which infiltrated into the beach during the flooding tide.

It seems apparent that the infiltrated saltwater is a major part of the shoreline seepage, which can vary widely, depending on the climatic conditions which enable fresh groundwater discharge, but perhaps more importantly on the beach face geometry and the magnitude of tide. While there is wide variability in estimates of fresh and saltwater discharge by the various approaches, it does indicate that a large proportion of beach discharge is infiltrated saltwater, probably in the range of 65–85% for the sites studied. In arid region coastlines it may be much higher, and in wetter coastal areas, much lower.   The distribution of subaerial and submarine seepage is more dependent on the beach characteristics of slope, hydraulic conductivity, and tidal range [Turner, 1995]. For Site A about 55% of the seepage is submarine seepage, but for Site B, only 35% is submarine seepage.

4.2.4   Shoreline Seepage Variability

In order to evaluate the physical evidence for variability of seepage along the beach front, soil samples were taken at both visually apparent high seepage zones and those that exhibited less seepage. It was found that the average median grain size for soil was 1.50 mm in the high seepage areas and 0.070 mm in the low seepage areas. It appears that once seepage is initiated it is self-enhancing by washing out the fines and creating higher hydraulic conductivity. Confirmation of this was established by employing Hazen's equation for hydraulic conductivity at the two zones which gave values of 78 m/day and 20 m/day for the high and low seepage zones, respectively. Airborne thermal infrared imagery was also used to ascertain the shoreline fresh groundwater discharge. This method is able to depict the freshwater discharge by imaging the temperature difference using spectral wave length differences between discharging fresh groundwater and the warmer receiving sea water [Portnoy et al., 1998] in the late summer.

As shown in Figure 9, for one of the study sites at Nauset Marsh, the freshwater discharge is shown as dark plumes emanating from the shoreline. Additionally, direct water quality measurements during low tide discharge at the site provided ground-truthing of the variation in shoreline salinity, as

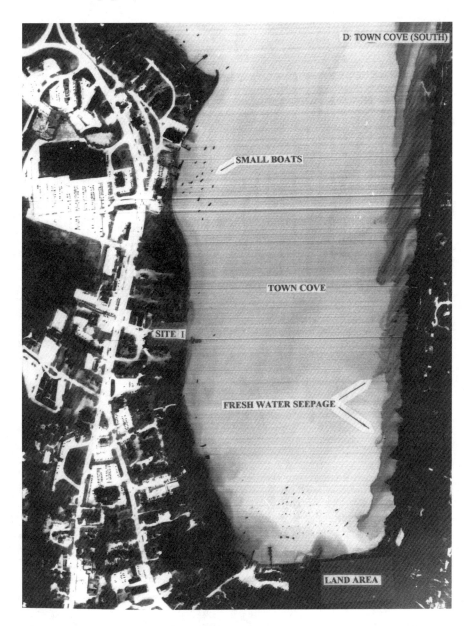

Figure 9: Thermal infrared imagery at Town Cove, Nauset Marsh.

illustrated in Figure 10. Dramatic indications of the shoreline variation in the salinity of discharge is evident, as well as the relationship of salinity with nitrogen, a selected chemical sampling parameter.

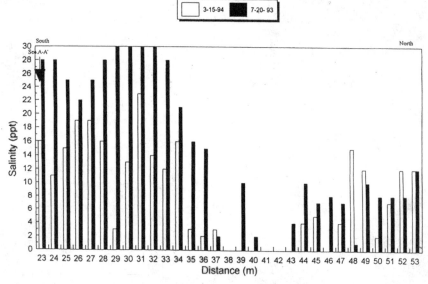

ELDREDGE SITE LOW TIDE SALINITY TRANSECTS

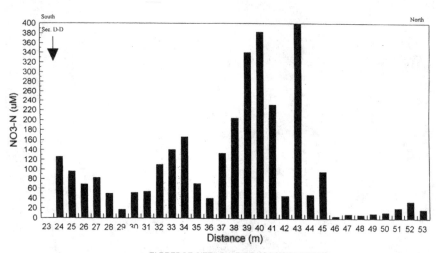

ELDREDGE SITE LOW TIDE NO3-N TRANSECT

Figure 10: Shoreline discharge salinity and nitrogen distribution at low tide.

5. SUMMARY

The nature of coastal groundwater seepage when viewed in the dynamic temporal and spatial context is highly complex, necessitating the use of many different methods and tools. It is best to begin with a broader based survey, such as the thermal infrared imagery or rapid shoreline salinity transects which will identify regions of fresh groundwater discharge. Then more focused attention can be effectively given to more detailed short-term and long-term sampling.

Acknowledgments

The information contained in the foregoing discussion is largely based on island and coastal groundwater studies funded by the National Science Foundation, Sea Grant, and the National Park Service. The support of these agencies is gratefully acknowledged as well as that of the many colleagues and graduate students who participated in the fieldwork.

REFERENCES

Baird, A.J. and Horn, D.P., "Monitoring and modeling groundwater behaviour in sand beaches", *Journal of Coastal Research*, **12**(3), 630–640, 1996.

Barwell, V.K. and Lee, D.R., "Determination of horizontal to vertical hydraulic conductivity ratios from seepage measurements on lake beds", *Water Resources Research*, **17**(3), 565–570, 1981.

Bokuniewicz, H., "Groundwater seepage into Great South Bay, New York", *Estuarine and Coastal Marine Science*, **10**(4), 437–444, 1980.

Bokuniewicz, H. and Pavlik, B., "Groundwater seepage along a barrier island", *Biogeochemistry*, 1990.

Bokuniewicz, H., "Analytical descriptions of subaqueous groundwater seepage", *Estuaries*, **15**(4), 458–464, 1992.

Cooper, H.H., Jr., Kohout, F.A., Henry, H.R., and Glover, R.E., "Sea water in coastal aquifers", U.S. Geological Survey Water-Supply Paper 1613-C, Washington, D.C., 82 p., 1964.

Drabbe, J. and Badon Ghijben, W., "Nota in verband met de voorgenomen putboring nabij," Amsterdam. Tsch. Kon. Inst. v. Ingenieurs, Verh. 1888–1889, The Hague. 8–22, 1889.

Frohlich, R.K., "Vertical electrical resistivity soundings in a Provincetown beach", personal communication, April, 2002.

Giblin, A.E. and Gaines A.G., "Nitrogen inputs to a marine embayment: the importance of groundwater", *Biogeochemistry*, **10**, 309–328, 1990.

Glover, R.E., "The pattern of fresh water flow in a coastal aquifer", *Journal of Geophysical Research*, **64**(4), 457–459, 1959.

Glover, R.E., "The Pattern of fresh water flow in a coastal aquifer", U. S. Geological Survey Water Supply Paper 1613C:C32–C35, 1964.

Hegge, B.J. and Masselink, G., "Groundwater-table responses to wave run-up: an experimental study from Western Australia," *Journal of Coastal Research*, **7**(3), 623–634, 1991.

Henry, H.R., "Interfaces between salt water and fresh water in coastal aquifers", U.S. Geological Survey Water Supply Paper 1613C: C35–C70, 1994.

Herzog, C., *Reed's Nautical Almanac, North American East Coast*, Thomas Reed Publications, Providence, RI, 2003.

Herzsberg, A., "Die Wasserversorgung," Jahrb. Jahrg., **44**, Munich. 815–819, 842–844, 1901.

Hubbert, M.K., " The theory of groundwater motion", *Jour. Geology*, **48**, no. 8, pt. 1, 785–944, 1940.

Kohout, F.A., "Fluctuations of ground-water levels caused by dispersion of salts", *Journal of Geophysical Research*, **66**(8), 2424–2434, 1961.

Lee, D.R., "A device for measuring seepage flux in lakes and estuaries", *Limnology and Oceanography*, **22**, 140–147, 1977.

Li, L., Barry, D.A., Stagnetti, F., Parlange, J.Y., Jeng, D.S., "Beach water table fluctuations due to spring-neap tides: moving boundary effects", *Advances in Water Resources*, **23**, 817–824, 2000.

Lusczynski, N.J., "Head and flow of ground water of variable density", *Journal of Geophysical Research*, **66**(12), 4247–4256, 1961.

McBride, M.S. and Pfannkuch, H.O., "The distribution of seepage within lake beds", Journal Research, U.S. Geological Survey 3, no. 5: 505–512, 1975.

Millham, N.P. and Howes, B.L., "Nutrient balance of a shallow coastal embayment: I. Patterns of groundwater discharge", Marine Ecology Progress Series 112, 155–167, 1964.

Nielsen, P., "Wave setup: a field study", *Journal of Geophysical Research*, **93**(C12), 15643–15652, 1988.

Nielsen, P., "Tidal dynamics of the water table in beaches", *Water Resources Research*, **26**(9), 2127–2134, 1990.

Portnoy, J.W., Nowicki, B.L., Roman, C.T., and Urish, D.W., "The discharge of nitrate-contaminated groundwater from developed shoreline to marsh-fringed estuary", *Water Resources Research*, **34**(11), 3095–3104, 1988.

Robinson, M.A. and Gallagher, D.L., "A model of groundwater discharge from an unconfined aquifer", *Ground Water*, **37**(1), 80–87, 1999.

Turner, I., "Water outcropping on macro-tidal beaches: a simulation model", *Marine Geology*, **115**, 227–238, 1993a.

Turner, I., "The Total Water Content of Sandy Beaches", *Journal of Coastal Research*, **S1**, no. 15, 11–26, 1993b.

Turner, I., "Simulating the influence of groundwater seepage on sediment transported by the sweep of the swash zone across micro-tidal beaches", *Marine Geology*, **125**, 153–174, 1995.

Urish, D.W., "Asymmetric variation of Ghyben-Herzberg lens", *Jour. Hydr. Div. Proc. American Society of Civil Engineers*, **107**, 1149–1158, 1980.

Urish, D.W., "The Effect of Beach Slope on the Fresh Water Lens in Small Oceanic Landmasses", Technical Completion Report, National Science Foundation Grant No. Eng-7908084, 1982.

Urish, D.W. and Ozbilgin, M., "The Coastal Ground-Water Boundary", *Ground Water*, **26**, 267–289, 1989.

Urish, D.W. and Qanbar, E., "Hydrologic Evaluation of Groundwater Discharge at Nauset Marsh, Cape Cod National Seashore, Massachusetts", Technical Report NPS/NESO-RNR/NRTR/97-07, 1997.

CHAPTER 8

Integrating Surface and Borehole Geophysics in the Characterization of Salinity in a Coastal Aquifer

F.L. Paillet

1. INTRODUCTION

In general, neither surface nor borehole geophysical methods can be used alone for the characterization of coastal aquifers *in situ*. This rather broad statement is based on the observation that surface geophysical surveys almost never have enough resolution to unambiguously define subsurface conditions [Sharma, 1997]. Much more definitive characterization can usually be performed using borehole geophysics, but there are never enough boreholes to effectively characterize complex formations on the basis of borehole data alone. Therefore, this discussion starts from the premise that effective characterization of subsurface hydrogeologic conditions in a heterogeneous coastal aquifer needs to be based on an effective integration of surface and borehole geophysics with other geologic and hydrogeologic data. At least in concept, subsurface characterization can be completed by using a limited set of borehole measurements to calibrate and otherwise condition a set of surface geophysical measurements that provide complete, three-dimensional coverage of the study region.

Although the need to combine surface and borehole geophysics in site characterization seems obvious, there are few published guidelines as to how to carry out such data integration. Some researchers recommend the "toolbox" approach where a variety of geophysical techniques (the tools) are considered, and a suite of the most appropriate kinds of measurement is used to complete characterization [Haeni *et al.*, 2001]. This study considers an analogous set of "conceptual tools" that might be used for the formulation of an effective and much less ambiguous joint integration of surface and borehole geophysics with other site data. We first list a number of such tools that might serve as a basis for the formulation of a geophysical data inversion and interpretation scheme. We then consider a large-scale site characterization study where each of these generalized conceptual tools was

1-56670-605-X/04/$0.00+$1.50
© 2004 by CRC Press LLC

applied to the data integration, and where the specific contribution of each can be identified. The results show that non-invasive characterization of heterogeneous coastal aquifers can be substantially improved by careful attention to the integration of surface and borehole geophysics during the course of the investigation.

2. THE CONCEPTUAL TOOLBOX

In analogy with the many different kinds of geophysical survey equipment available, there are a number of basic concepts that can be applied to the inversion of geophysical data regardless of whether that data is electric, acoustic, or some other class of physical measurement. Because these "conceptual tools" offer a general way of interpreting almost any kind of geophysical data, they can be considered in formulating almost any subsurface investigation. We find a set of five such tools that could, in theory, be applied to any geophysical study in general, and coastal aquifers in particular.

2.1 Scale of Investigation

Any geophysical survey made at the surface of the earth can, in principle, be made over a much smaller scale of investigation in a borehole. This concept allows the direct investigation of scale of measurement on geophysical response (Figure 1). The surface surveys average measurements over progressively larger sample volumes (defined by R_1, R_2, etc. in Figure 1) as the depth of investigation is increased. The borehole log makes the same measurement (electrical induction, acoustic velocity, bulk density, etc.) over a small sample volume (defined by R_0 in Figure 1) as the probe is moved along the borehole. Thus, we have a means to investigate how small sub-samples within the surface survey volume contribute to the larger-scale geophysical response of the formation.

2.2 Regression of Borehole Data to Calibrate Surface Measurements

The exact relation between surface geophysical surveys and hydraulic or geologic properties of interest in the subsurface is often not well known. Because the same kind of measurement can also be made in the borehole over a smaller sample volume, geophysical logs provide for direct regression of a geophysical measurement with aquifer parameters given by hydraulic tests or water sample analyses. In this approach, the geophysical log response can be averaged over the screened interval in a test well (Figure 1) and this value can be used to calibrate the surface survey in terms of the hydraulic or water quality property of interest in a particular study. A typical

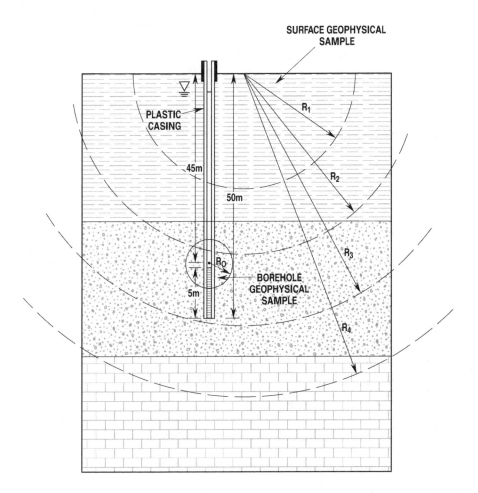

Figure 1: Schematic illustration of scales associated with surface and borehole geophysical measurements compared to typical screened interval for hydraulic testing and water sample analysis.

example is given in Figure 2, where the induction log measurement of formation conductivity is averaged over the screened interval in a sampling well to construct a relation between the electrical conductivity of the formation and electrical conductivity (salinity) of the water sample.

2.3   Multivariate Interpretation From Standard Logs

Almost all geophysical properties that can be measured at the surface are a function of more than one subsurface variable. Given that fact, a single surface survey cannot be effectively related to one variable of interest where

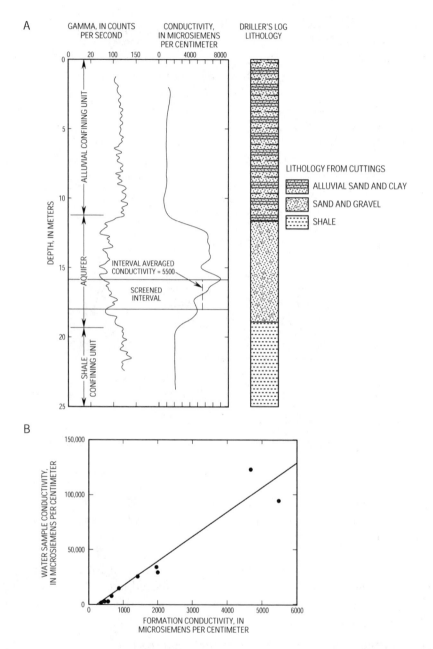

Figure 2: Example of borehole induction methods used to develop a regression between water quality and formation resistivity: A) formation conductivity averaged over screened interval in a sampling well; and B) regression of electrical conductivity of water sample to formation conductivity for a series of monitoring wells.

there is significant variation related to other properties of the subsurface. Geophysical soundings in coastal aquifers are most often made with electromagnetic methods to identify the relatively high electrical conductivity of sediments saturated with saline water. However, the electrical conductivity of porous material is determined by several different factors, such as the solute content of pore water, the electrical conductivity of the mineral matrix, and the geometry of pore spaces. Therefore, subsurface pore-water salinity cannot be uniquely determined on the basis of the measurement of subsurface electromagnetic properties alone. Several different geophysical logs can be run in boreholes and can be interpreted to define a physical model for the multivariate properties of the subsurface. In Figure 2, one geophysical log (natural gamma log) is used to define the aquifer. The combination of gamma and induction logs shows that formation electrical conductivity depends on both pore-water conductivity (in the aquifer) and on formation lithology (clay minerals in the overlying clay-rich alluvium, and in the underlying shale). In this example, the logs demonstrate that the regression between water conductivity and formation electrical conductivity can only be used where the surface geophysical survey interpretations apply to the sand and gravel aquifer. Effective interpretation of surface electromagnetic surveys in terms of water conductivity will only result when either surveys affected by the electrical conductivity of clay minerals in the surficial alluvium are removed from the data set or an interpretation model is used to account for the presence of this alluvium.

2.4   Inversion Model Characteristics in Data Inversion

The mathematical challenge of geophysical data inversion usually comes down to relating a finite number of surveys to a continuous distribution of subsurface properties. No matter whether the inversion involves one, two, or three dimensions, the continuous distribution in each dimension can be approximated as a series expansion [Parker, 1994]. There are an infinite number of coefficients in each such expansion. Thus, we never have enough data to form a series of equations relating the finite measurements to the infinite unknown coefficients. One solution is to truncate the series expansions to fewer coefficients than there are data points. This means that there are more equations than unknowns, and the residuals from the additional equations can be used to reduce the mean square difference between model and data. That is, the various empirical parameters used in the inversion model can be systematically adjusted to find a solution where there is a minimum residual error when the solution is substituted in the full set of inversion equations. Geophysical logs provide information about the actual distribution of properties in the subsurface that can be used to determine how many coefficients to retain in the expansion, or which series of basis functions to use.

In the practical application of inversion algorithms developed for each class of surface geophysical survey, the user can determine the number of subsurface layers or cells to be used in the analysis. There will always be a reduction in the residual error of the best-fit solution as the number of layers or cells is increased. The geophysicist has to decide whether the improvement in the fit of the model to the data set is offset by the reduction of degrees of freedom in the analysis. There are quantitative statistical tests that can be applied to determine whether the improvement is statistically significant, but such tests generally require knowledge of the statistical properties of the subsurface. The specific information about the subsurface structure provided by geophysical logs can significantly improve the ability to formulate and interpret geophysical inversion problems. It is also known that certain geophysical measurements cannot distinguish between alternate subsurface models (for example, electrical equivalence) [Sharma, 1997]. Geophysical logs can provide the information needed to resolve the ambivalences inherent in the selection of a specified inversion model from among equivalent models.

2.5   Verification Boreholes

When geophysical surveys are interpreted, the final analysis of the data set gives a prediction of subsurface properties over regions between boreholes. A statistically significant verification of the model can be obtained by identifying regions where the model predicts specific features, such as the center of a buried valley or a sharp contrast in the salinity of pore water. Geophysical logs in verification boreholes, commonly drilled at minimal expense by standard rotary drilling, then left uncased and kept open with drilling mud, can be used to verify that these features are present as predicted. When logs show that features predicted by the model actually exist, the results provide almost irrefutable evidence in support of the interpretation. Considerable care can be taken to ensure that the verification boreholes are drilled in locations that effectively test the inversion model predictions, so as to maximize the impact of model verification.

3.   THE SOUTH FLORIDA STUDY

Surface and borehole geophysics were combined with core descriptions, water sample analyses, and hydraulic tests to generate a predictive model for the surficial aquifers in the region surrounding the Big Cypress National Preserve in south Florida [Weedman et al., 1997; Bennet,

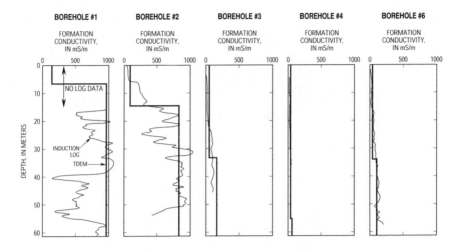

Figure 3: Surface time-domain electromagnetic (TDEM) surveys at borehole sites demonstrate that surveys define the electrical conductivity of the uppermost layer, and the composite electrical conductivity of underlying aquifers and confining units.

1992]. In this study, surface time-domain electromagnetic surveys (TDEM) [Fitterman and Stewart, 1986; Kaufman and Keller, 1983] were used to project aquifer structure and water quality conditions identified at individual boreholes over the more than 10 km distances between individual drilling sites. The south Florida geophysical data analysis provides a useful example of the contribution of borehole geophysics to the interpretation of surface geophysical surveys [Paillet *et al.*, 1999; Paillet and Reese, 2000]. In the following sections, each of the conceptual interpretation tools described above is evaluated with respect to its contribution to the electromagnetic survey example from south Florida.

3.1 Scale of Investigation

Because the focus of the south Florida study was water quality and possible seawater intrusion, the electrical conductivity of the surficial aquifer was of primary interest. The relationship between electrical conductivity and formation properties could be compared at both geophysical log and surface survey scales of investigation (Figure 3). Although other information would be required to generate a useful model of subsurface properties on the basis of this combination of data, the comparison of electrical conductivity measured at two such very different scales of investigation confirms that the local variations of induction conductivity can be related to the

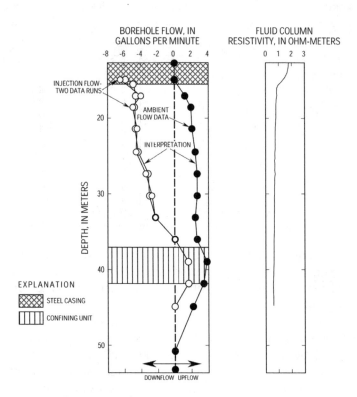

Figure 4: Flow logs obtained in fully screened boreholes showed that natural upward flow existed in most boreholes; the fluid column resistivity profiles of boreholes under ambient conditions could then be unambiguously related to the electrical conductivity of pore water in the inflow zone or zones.

depth-averaged measurements of subsurface conductivity given by the surface surveys. For this reason, the comparison of surface and borehole measurements of formation electrical conductivity served as an ideal starting point in the construction of a valid inversion model for the surface TDEM surveys.

3.2   Water Quality Regression

In the south Florida study, it was possible to relate water quality in a number of zones to local formation conductivity because there was natural flow in most boreholes after completion by installing fully screened casing and flushing of drilling mud (Figure 4). Under those flow conditions, the fluid column resistivity (0.8 ohm•m in Figure 4) could be unambiguously related to the electrical conductivity (12,500 µS/cm) of the pore water entering the borehole in the inflow interval (45–52 m in Figure 4). This analysis was repeated in all boreholes where natural flow was present, and where inflowing water ranged in conductivity from less than 1000 μS/cm to more than 14,000 μS/cm. The regression between formation electrical conductivity and pore water conductivity generated a slope of about 2.3 (the formation factor) that could be used to relate formation electrical conductivity to pore water electrical conductivity in all geophysical measurements where pore water conductivity could not otherwise be determined. This formation factor appeared anomalously low as compared to typical values of greater than 20 in consolidated sandstone [Hearst *et al.*, 2000] but was attributed to the association of inflow with the most permeable intervals within aquifer units characterized by unusually high transmissivity values as reported by Paillet and Reese [2000]. Empirical studies demonstrate that the formation factor decreases as permeability increases [Biella *et al.*, 1983; Jorgensen, 1991].

3.3   Multivariate Dependence of Formation Properties

In general, formation electrical conductivity depends on the salinity of pore water, the influence of pore network geometry (permeability), ion mobility, and the fraction of electrically conductive minerals (clays) [Biella *et al.*, 1983; Jorgensen, 1991; Kwader, 1985]. Geophysical logs from the south Florida boreholes indicated that the contribution of lithology and pore structure to variations in electrical conductivity were negligible (Figure 5). Although the erratic distribution of phosphatic sands caused gamma logs to be of no use in characterizing these sediments, comparison of neutron and induction logs with core lithology confirmed that clays were absent and that formation electrical conductivity and porosity trends ran parallel over discrete intervals [Weedman *et al.*, 1997]. This result indicates that the subsurface at each borehole site consists of a series of aquifer layers, each characterized by a single value of pore water salinity. Thin confining units separate aquifers of different pore water salinity, accounting for the step-wise increase in subsurface electrical conductivity. These results indicate that an effective large-scale model for the surficial aquifer is a series of aquifers of different thickness and containing water of differing salinity separated by thin, mineralized confining units.

3.4   Aquifer Structure and Inversion Layers

The layered aquifer framework interpreted from Figure 5 defines the surface electrical survey interpretation as the mapping of the aquifers and confining units identified at each borehole site over the distance between boreholes at this study site. The subsurface structure clearly indicates that

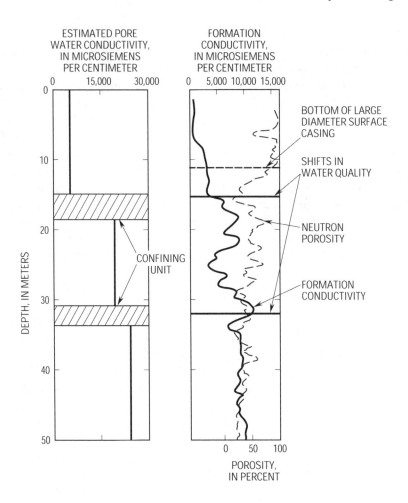

Figure 5: Overlay of induction and neutron porosity logs demonstrate that the surficial aquifer separated by thin confining units into aquifers containing pore water of different salinity, and suitable for electrical modeling as a layered system.

model inversion formulated as a series of layers is appropriate for this situation. The comparison of logs and surveys in Figure 3 shows that the surveys effectively indicate the electrical conductivity of the uppermost aquifer layer and the depth-averaged conductivity of the series of aquifers and confining units under that uppermost layer.

An example of the aquifer inversion model constructed from the TDEM surveys along a profile between two of the boreholes at the study site is given in Figure 6. The profiles show that the inversion can be completed

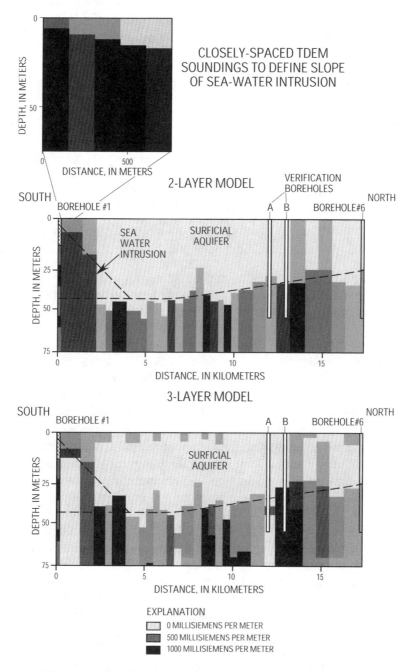

Figure 6: Time-domain electromagnetic survey profile across the study area for two-layer and three-layer inversion models, showing that there is no meaningful difference between the two-layer and three-layer inversions.

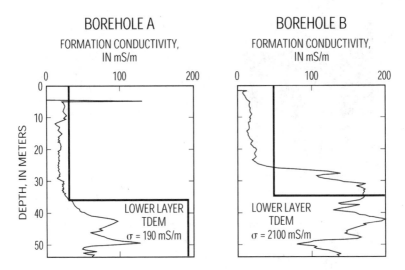

Figure 7: Induction log in rotary-drilled boreholes used to verify model predictions given in Figure 6A compared to the time-domain electromagnetic (TDEM) survey inversion at that location in the profile.

using two- or three-model layers for each survey, but that the profile constructed from either set of inversions shows the same structure. The profile indicates an unconfined surficial aquifer with the same water quality extends across the study site (pore-water electrical conductivity of about 400 μS/cm, and identical with the quality of the overlying surface water bodies). The underlying layer is interpreted as a composite of one or more confining units and the underlying aquifers. A wedge of seawater intrusion is interpreted on the southern side of the profile in Figure 6, corresponding with the landward limit of tidal fluctuation in local estuaries. Although not well resolved in this representation, the slope of this interface was verified by a series of more closely spaced TDEM surveys at the southern end of the TDEM profile as shown in Figure 6 [Paillet *et al.*, 1999]. Otherwise, the combination of randomly varying conductivity in the lower interpretation layer (two-layer model) and the presence of a strong upward hydraulic gradient throughout the study area indicates that subsurface variations in salinity are related to variations in the rate of upward seepage of brine and the local intrusion of seawater in the immediate vicinity of the coast.

3.5   Verification Boreholes

Although the interpretation based on data such as those shown in Figure 6 appears convincing, a few additional boreholes drilled at carefully selected locations would strongly support the interpretation if they showed the depth to the bottom of the upper aquifer and the electrical conductivity of the underlying sediments were in agreement with prediction. Three such boreholes were drilled as part of the south Florida study; two of these are indicated in Figure 6. Induction logs from those boreholes agreed quantitatively with the predictions, showing a definite step in conductivity at the predicted depth and the predicted depth-averaged electrical conductivity of the upper aquifer (Figure 7). The agreement for the underlying zone was only qualitative in that the logs showed an electrical conductivity for the lower layer significantly less than predicted, but the relative magnitude of the measured conductivity agreed with the predictions. That is, the logs in Figure 7 show that the underlying layer is significantly less conductive in borehole A than in borehole B, as indicated by the TDEM surveys. One important result is that the depth to the interface given by the logs in the verification boreholes corresponded with the average depth of the interface constructed from the average of several adjacent TDEM stations (the dashed line in Figure 6), and not the interface given by the TDEM station nearest to the drilling site. This result provides concrete justification for the otherwise reasonable but unproved assumption that the bottom of the surficial aquifer is given by the average trend of the TDEM surveys in the profile.

4. CONCLUSIONS

Integration of geophysical data obtained at various scales provides an effective way to bridge the gap between localized data from boreholes and site-wide data from regional surveys of coastal aquifers. Specific conceptual approaches to such analysis are summarized in Table 1. The contribution of each of these approaches to multiple-scale geophysical site characterization was assessed at a study site in south Florida. Comparison of induction logs in boreholes with surface time-domain electromagnetic surveys near borehole locations was critical in developing a model relating aquifer framework, water quality, and large-scale electrical conductivity layers for the interpretation model. Regression of pore water electrical conductivity measured in boreholes against formation conductivity given by induction logs was effectively used to interpret the salinity of pore water in the upper aquifer layer using the surface surveys. Joint analysis of the combination of surface electromagnetic surveys and borehole geophysical logs indicated that salinity of water in the surficial aquifers at the study site is controlled by a simple wedge of seawater intrusion along the coast and drilling by a complex pattern of upward brine seepage from deeper aquifers throughout the study site. This interpretation was independently checked by three test boreholes to verify the location of aquifer boundaries and the relative salinity of subsurface waters as predicted by the analysis.

Conceptual Tool	Geophysical Measurement	Example of Application
Scale of investigation	Vertical distribution of electrical conductivity	Compare number and thickness of layers used in inversion of electromagnetic soundings with local structure given by induction log
Regression of data	Electrical conductivity of formation around the borehole	Compare induction log to water sample data for screened intervals to calibrate electric surveys in water-quality units
Multivariate interpretation	Electrical conductivity of water-bearing sediments	Use logs to identify lithology to distinguish effects of mineral matrix conductivity from the effects of pore water salinity on formation electrical conductivity
Inversion methods	Inversion model structure and minimizing residuals	Compare logs to results of inversion models to define optimum size and shape of model layers or cells
Verification boreholes	Subsurface structure from surface soundings	Log rotary-drilled boreholes to determine local depth and thickness of geo-electric layers as inferred from electromagnetic sounding interpretation

Note: For additional details on surface and borehole geophysics in environmental and ground water applications consult Paillet and Crowder [1996] and Sharma [1997]. Readers are also referred to American Society for Testing and Materials, Standard Guide for Selecting Surface Geophysical Methods, (D 6429-99).

Table 1: Conceptual tools used in the integration of surface and borehole geophysics.

REFERENCES

Bennet, M.W., A Three Dimensional Finite Difference Ground Water Flow Model of Western Collier County, Florida, Technical Publication 92-04, South Florida Water Management District, West Palm Beach, Florida, 1992.

Biella, G., Lozeij, A., and Tabacco, I., "Experimental study of some hydrogeological properties of unconsolidated porous media," *Ground Water*, **21**, 741, 1983.

Fitterman, D.V., and Stewart, M.T., "Transient electromagnetic sounding for groundwater," *Geophysics*, **51**, 995, 1986.

Haeni, F.P., Lane, J.W. Jr., Williams, J.W., and Johnson, C.D., "Use of a geophysical toolbox to characterize ground-water flow in fractured rock," in *Proc. Fractured Rock 2001 Conference*, Toronto, Ontario, March 26–28, 2001, Smithville Phase IV Bedrock Remediation Program, CD-ROM.

Hearst, J.R., Nelson, P.H., and Paillet, F.L., *Well Logging for Physical Properties*, 2nd ed., John Wiley and Sons, Ltd., New York, 2000.

Jorgensen, D.G., "Estimating geohydrologic properties from borehole geophysical logs, *Ground Water Monitoring and Remediation*, **10**, 123, 1991.

Kaufman, A.A., and Keller, G.V., *Frequency and Transient Soundings*, Elsevier Science Publishers, Amsterdam, 1983.

Kwader, T., "Estimating aquifer permeability from formation resistivity factor," *Ground Water*, **23**, 762, 1985.

Paillet, F.L., and Crowder, R.E., "A generalized approach for the interpretation of geophysical well logs in ground water studies— Theory and application," *Ground Water*, **34**, 883, 1996.

Paillet, F.L., Hite, L., and Carlson, M., "Integrating surface and borehole geophysics in ground water studies—An example using electromagnetic soundings in South Florida," *Journal of Environmental and Engineering Geophysics*, **4**, 45, 1999.

Paillet, F.L., and Reese, R.S., "Integrating borehole logs and aquifer tests in aquifer characterization," *Ground Water*, **38**, 713, 2000.

Parker, R.L., *Geophysical Inverse Theory*, Princeton University Press, Princeton, New Jersey, 1994.

Sharma, P.V., *Environmental and Engineering Geophysics*, Cambridge University Press, Cambridge, UK, 1997.

Weedman, S.D., Paillet, F.L., Means, G.H., and Scott, T.M., Lithology and Geophysics of the Surficial Aquifer System in Western Collier County, Florida, Open-File Report 97-436, U.S. Geological Survey, Reston, Virginia, 1997.

CHAPTER 9

Geographical Information Systems and Modeling of Saltwater Intrusion in the Capoterra Alluvial Plain (Sardinia, Italy)

G. Barrocu, M.G. Sciabica, L. Muscas

1. INTRODUCTION

A comprehensive study of the Capoterra alluvial plain (Southern Sardinia, Italy) has been carried out by the Engineering Geology and Applied Geophysics Section of the Department of Land Engineering at Cagliari University, within the frame of the international projects MEDALUS and AVICENNE 73, funded by the European Union. The purpose of the investigation was to gain a deeper insight into saltwater intrusion in the coastal aquifer system and to numerically simulate the phenomenon [Sciabica, 1994; Barrocu *et al.*, 1997; Barrocu *et al.*, 1998].

On account of the large amount of data collected for the Capoterra alluvial plain, a Geographical Information Systems (GIS) was set up, creating an alphanumerical database together with a geographic database, so as to enable integrated methods to be adopted for modeling saltwater intrusion in the coastal aquifer system.

The comprehensive study consisted of the following main phases:

- set up of a control and monitoring network of the coastal aquifer system;
- hydrogeological and hydrogeochemical measurements at selected observation wells in the network;
- pumping tests and artificial recharge experiments at selected observation wells or purposely built wells and piezometers;
- definition of the hydrogeological model of the alluvial plain;
- development of a modeling procedure for simulating saltwater intrusion phenomena in the coastal aquifer system;
- organization of the data collected into a GIS for modeling saltwater intrusion.

The main purposes of the study were:

1-56670-605-X/04/$0.00+$1.50
© 2004 by CRC Press LLC

- to define the hydrogeological model of the alluvial plain;
- to determine the hydrogeological and physical-chemical parameters of the aquifer system;
- to build an alphanumeric database;
- to validate saltwater intrusion modeling results;
- to apply GIS and modeling procedures as integrated methods for studying saltwater intrusion in coastal aquifers.

The scheme of the comprehensive study carried out in the Capoterra alluvial plain and its main objectives are shown in Figure 1. Fieldwork, field data collection, and processing phases are indicated in green, study activities and modeling procedure in blue, the specific objectives in red.

Since June 1991, hydrogeological and hydrogeochemical measurements have been taken in the control and monitoring network set up in the plain. The complete set of data forms an alphanumeric database, which is constantly updated and used for constructing graphical representations of the hydrogeological and hydrogeochemical data. Pumping tests and artificial recharge experiments have been performed at selected observation wells or at purposely built wells and piezometers to determine the parameters for simulating saltwater intrusion.

The purpose of the hydrogeological and hydrogeochemical study was to refine the hydrogeological model for improving the saltwater intrusion modeling validation. The information collected during the investigation carried out in the Capoterra alluvial plain was used to build a GIS to aid saltwater intrusion modeling.

The investigation had three main objectives: firstly to develop a hydrogeological model of saltwater intrusion in the alluvial plain, secondly to validate the modeling procedure, and lastly, more in general, to apply the GIS and the modeling procedure as integrated methods for studying saltwater intrusion in coastal aquifers, for the purpose of defining strategies for managing integrated resources.

2. PHYSIOGRAPHY, SURFACE AND GROUNDWATER HYDROGEOLOGY

2.1 Field Investigations and Monitoring Network

The Capoterra alluvial plain is situated in the southwestern portion of the Campidano Graben in southern Sardinia (Italy). It comprises, to the south, the delta of the Santa Lucia River, a torrential watercourse, and is bounded eastward by the Santa Gilla lagoon and northward by the Cixerri River. To the west it is interrupted by a series of hills aligned *en échelon,*

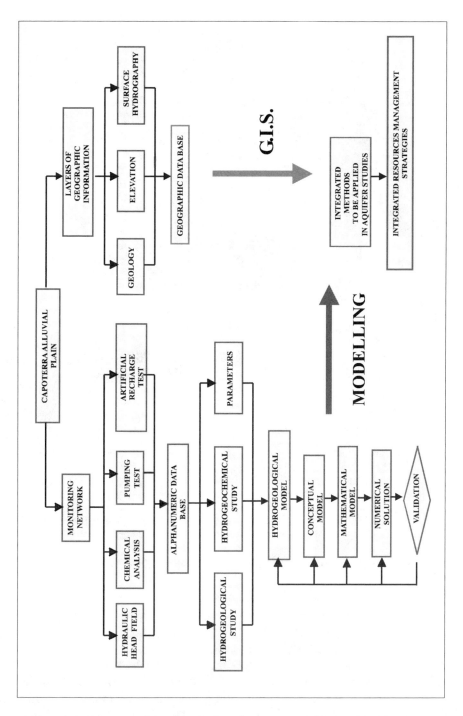

Figure 1: Scheme of the comprehensive study and its main objectives.

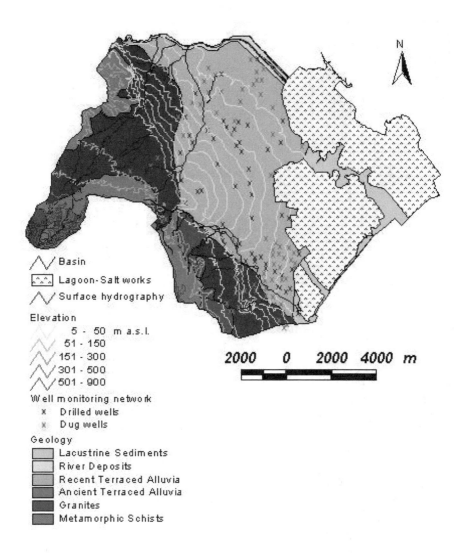

Figure 2: Capoterra alluvial plain (Southern Sardinia, Italy).

representing the extension of the tectonic block that west of the Sardinian Graben is split up by two main sets of NW–SE and NE–SW trending fractures (Figure 2). (See the accompanying CD for color figures.)

Based on the geological, hydrogeological, geomorphological, and pedological information for the area, the following hydrogeological units have been recognized [Barrocu et al., 2000]: fluvial and lacustrine sediments, recent and ancient terraced alluvia of the Quaternary, and fractured granites and metamorphic schists of the Palaeozoic. The recent alluvia are highly permeable and contain a phreatic aquifer, overlaying a second multi-layer aquifer, semi or locally confined.

Over the last two decades, the area has undergone profound transformations due to agricultural and industrial expansion, and water demand has increased accordingly. The particular climatic conditions of the area, characterized by frequent and prolonged periods of drought as well as the presence of a variety of natural and anthropic sources of salt (sea spray, sea water, lagoon, evaporation ponds of the salt-works and salt-hills), combined with indiscriminate overexploitation have resulted in the depletion of groundwater resources and in their widespread salination, with more serious effects in the shallow part of the coastal aquifer system.

A monitoring network for controlling water quality and groundwater level was set up in June 1991, initially in the southern portion of the plain, near the Santa Lucia River, and then extended northward in April 1992 so as to include the area nearer the Cixerri River. The network consists of 132 wells, 74 wide-diameter and relatively shallow wells, dug into the superficial aquifer, and 58 wells drilled into the deep aquifer.

In 1991, 1992, and 1993, groundwater levels were measured each month in all the wells, and water samples collected from some significant wells were chemically analyzed in the laboratory. In April 1994 chemical determinations were done on samples of water taken from the sea, from three evaporation stages of the saltpans, from the lagoon and from the Rio Santa Lucia and Rio Cixerri.

Pumping tests were performed at selected observation wells or at purposely built wells and piezometers so as to evaluate the hydrogeological parameters for simulating saltwater intrusion. Artificial recharge experiments were also carried out at purposely built wells and piezometers in the plain so as to assess the efficiency of a hydrodynamic barrier aimed at controlling saltwater encroachment and its spatial evolution [Barrocu et al., 1997]. More recently, in July 1998 a measurement campaign was conducted in the frame of a detailed study of the groundwater geology and geochemistry [Vernier, 1999].

2.2   Hydrogeological and Hydrogeochemical Investigation Results

The hydrogeological and hydrogeochemical data, collected during measurements taken in the control and monitoring network, were used for constructing graphic representations such as piezometric contour lines, Schoeller's diagram, Chebotarev's diagram, TDS (total dissolved solid) contour lines, and so on [Barrocu et al., 1994; Barrocu et al., 1994].

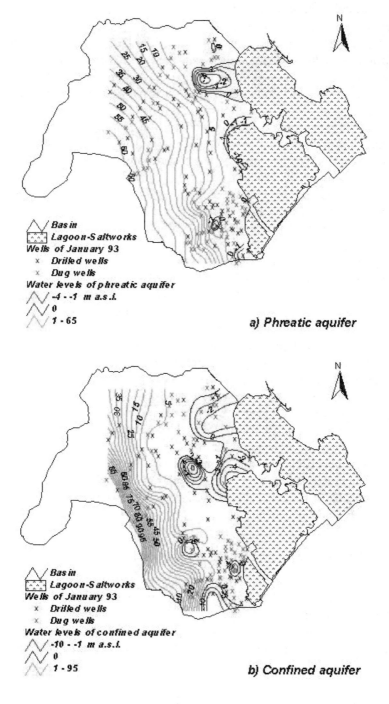

Figure 3: Water level contour lines measured in January 1993.

The water level contour lines for January 1993 (Figure 3) show that both aquifers are recharged laterally by groundwater through the granite bedrock at the western boundary of the aquifer. Supply probably takes place through preferential pathways in the form of the main fracture systems occurring in the bedrock. In both aquifers there is a depression of piezometric surface to below mean sea level coinciding with groundwater over-exploitation to satisfy agricultural and industrial demand. Depressions are located mainly in the central part of the plain, near to the lagoon and the saltworks. Low piezometric levels were also observed in the proximity of the coast.

The surface aquifer is generally exploited for irrigation during the hot and dry summer months. Drawdown coincides with two major abstraction areas, one to the NE of the plain, near the lagoon, the other to the SE, in the vicinity of the saltworks. In winter, the irrigation demand decreases significantly and drawdown disappears owing to infiltrating precipitation and lateral recharge of the aquifer. Inspection of the contour lines in the dry summer months and in the rainy winter period clearly shows that the zero drawdown contour line migrates from inland to the edge of the lagoon (see the accompanying CD).

The drawdown in the confined aquifer, which varies over time depending on precipitation, is located for the most part in the central portion of the plain where a number of wells have been drilled for supplying industrial water (see the accompanying CD).

Figure 4 shows the January 1993 TDS contour lines for both aquifers. Recharge into the phreatic aquifer from the western side is indicated by the lowest salt content; to the south (coastal zone) and to the east (where the saltworks and lagoon are located) the values increase, corresponding to the lowest piezometric heads.

The saltworks constitute a possible source of salination of shallow groundwater. The salt froth that forms in the evaporation ponds during the various processing stages, along with the salt produced and stored in salthills in the open air, are scattered over the plain by the strong winds typical of the area. Assuming this to hold true, salination in the eastern portion of the plain and near the border with the saltworks remains practically unchanged throughout the year. In summer the situation near the coast is exacerbated by the increase in irrigation demand.

In the main, the deep aquifer is not as salinated as the phreatic one. In the central part of the plain, where intense groundwater abstraction results in significant depression of the piezometric surface, salinity remains low all year round, while in the western part of the plain it exhibits an increasing trend. Two hypotheses may be advanced to explain this phenomenon:

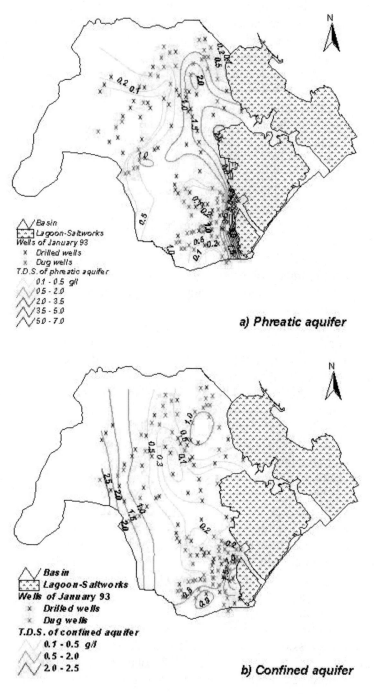

Figure 4: TDS contour lines measured in January 1993.

- two different layers are present in the confined aquifer: the upper layer, near the edge of the plain, is recharged slowly through the granite rocks, while recharge of the second layer probably occurs through preferential pathways via a deeper and faster circuit through the granite bedrock;
- connate saltwater may be entrapped in old unleached sea deposits between the fractures controlling lateral inflow into the confined aquifer and have probably been extracted by pumping.

Piper's diagram constructed on the basis of the chemical composition of all water sampled from the two aquifers has shown the groundwater in the plain to have similar composition, in general of the alkaline-chloride-sulphate type. Based on Stuyfzand's classification practically all groundwater can be classified as the NaCl type.

The fresh water end-member for each aquifer has been chosen comparing the chemical composition of the fresher shallow or deep groundwater with granite spring water: all waters are of the NaCl type and have very low TDS. Typically groundwater hosted in granite rocks is not of the NaCl type: in the case at hand, however, the salt from evaporated sea-spray deposited on the granite hills is dissolved by atmospheric water.

Starting with the composition of the freshwater end-members selected for the two aquifers, many different processes take place to alter the original chemical composition according to the observed scattering of points in the various diagrams constructed for the hydrogeochemical study. These include carbonate dissolution and/or gypsum solution, Na/Ca or Ca/Na exchange, calcium-sulphate dissolution. The superposition of these different processes makes the situation very complex [Barrocu *et al.*, 1994] (for more details see the accompanying CD).

3. MODELING SALTWATER INTRUSION

3.1 Description of the Modeling Procedure

A groundwater model is a simplified representation of a real groundwater system or process. It is possible to define several types of groundwater models:

- The hydrogeological model is the collection of information describing the physical and human reality of a particular field area [Issar and Passchier, 1992].
- The conceptual model is the set of assumptions selected for verbally describing the processes that take place in the area [Bear and Bachmat, 1990].

- The mathematical model replaces the conceptual assumptions by mathematical expressions containing variables, parameters, and constants [Bear and Bachmat, 1990].

Discrepancies between observed and calculated data for a groundwater system are indicative of errors in the modeling procedure. There are several sources of errors:

- the hydrogeological model used to represent the particular field area;
- the conceptual model and its translation into the mathematical model;
- the numerical solution; and
- uncertainties and inadequacies in the input data that reflect our inability to describe the aquifer properties, stresses, and boundaries.

The most common sources of error in groundwater modeling lie in the conceptualization of the model and uncertainty of the data. Furthermore, the coexistence of several sources of error means that it may not be possible to distinguish among them.

The modeling procedure applied for simulating saltwater intrusion processes in the Capoterra coastal aquifer system involves the definition of the hydrogeological and conceptual models, formulation of the mathematical model, its numerical solution, and validation using field measurements and chemical determinations [Sciabica, 1994]. If the model is not satisfactorily validated, adjustments will have to be made at the various modeling stages and further simulation performed, repeating this procedure until the model has been successfully validated (Figure 5).

3.2   The Hydrogeological Model

The hydrogeological model has been defined by identifying both the natural factors, such as geology, proximity of the sea, presence of the lagoon and saltworks, and anthropic factors including irrigated agriculture and the urban and industrial development of the study area. The relationship between groundwater, lagoon, saltworks and sea has been established by processing piezometric data. Processes influencing water salinity in the coastal aquifer system were elucidated by examining the correlation between chemical elements present in the water samples.

Figure 6 shows a schematic representation of the hydrogeological model; the main elements are:

1. The coastal aquifer system consists of a phreatic aquifer and a semi or locally confined multilayer aquifer.
2. The two aquifers are interconnected in several places owing to numerous shoddily built wells with the result that the groundwater is locally mixed.

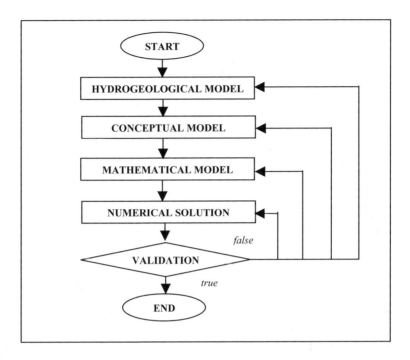

Figure 5: Schematic modeling procedure.

3. Both phreatic and confined aquifers are laterally recharged through the fractures of the granite bedrock on the west side of the plain, and they do not receive any water from the Cixerri River basin.

4. The confined multilayer aquifer probably contains two different aquifer levels, each supplied in a different way by waters flowing through the massive granite-metamorphic rocks.

5. Depressions of piezometric surface to below mean sea level coincide with groundwater over-exploitation to satisfy agricultural and industrial demand.

6. The reference freshwaters of the aquifers are chemically similar and are closely related to the groundwater of the granite bedrock.

7. Windblown salt from sea spray deposits considerable amounts of sodium and chlorine on the soils in the plain, altering the quality of the supply waters. Sodium and chlorine are also deposited by precipitation.

8. Due to the vicinity of the sea, seawater intrudes into the groundwater in the phreatic and confined aquifers of the plain, increasing the salt content.

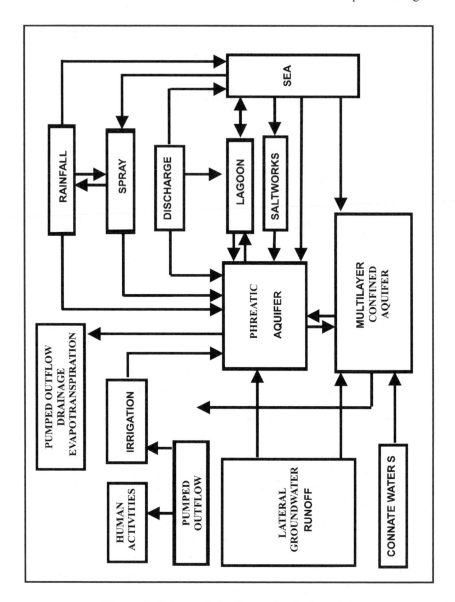

Figure 6: Schematic hydrogeological model.

9. The presence of the saltworks gives rise to additional phenomena other than seawater intrusion: windborne salt from the salt ponds and stocked in salt hills inside the saltworks near the evaporation ponds deposit on the soil surface.

10. Results of chemical analysis have shown a higher increase in salinity in the phreatic aquifer than in the confined one.

11. In the confined aquifer salinity is due to seawater encroachment produced by pumping to satisfy agricultural and industrial water demand.
12. In the confined aquifer connate waters, probably trapped in old unleached seawater deposits between flux cones departing from the fractures controlling lateral inflow into the aquifers, have been tapped by pumping.
13. In the phreatic aquifer salinity is caused by:
    - Brackish water encroachment from the saltworks and the lagoon.
    - Direct infiltration of the salts conveyed as windborne spray from the saltworks; these salts deposit on the soil and are then carried into the phreatic aquifer by infiltrating rainfall and irrigation waters.
    - Seawater intrusion.
14. During the summer irrigation waters dissolve the salts deposited on the soil.
15. In the dry season, when irrigation demand increases significantly, the salt concentration of the phreatic aquifer diminishes due to dilution by irrigation waters tapped through drilled wells from the deep aquifer, which has lower salinity than the shallow aquifer.
16. As a result, in the dry season irrigation water is recycled.

3.3   The Conceptual Model

The conceptual model has been defined simplifying the domain and the phenomenon under study, leading to the following assumptions:

A_1 The model is three-dimensional.

A_2 A single fluid phase (water) occupies the entire void space with only a single component (salt) completely dissolved in the water.

A_3 The freshwater–saltwater mixture is incompressible.

A_4 The salt is conservative.

A_5 Conditions are isothermal.

A_6 Viscosity does not affect the permeability of the porous medium.

A_7 The salt does not react with the porous matrix.

A_8 The salt concentration is not high enough to invalidate Darcy's and Fick's laws.

A_9 The density of the freshwater–saltwater mixture depends only on the salt concentration.

A_{10} The entire aquifer system is approximated by a phreatic aquifer of uniform thickness.

A_{11} The aquifer thickness is divided into five layers.

A_{12} Four homogeneous and isotropic zones are defined: lacustrine sediments, recent and ancient terraced alluvia of the Quaternary, fractured granites, and metamorphic schists of the Palaeozoic.

A_{13} There is no artificial recharge (irrigation) or natural replenishment (precipitation).

A_{14} Boundary conditions for flow and transport equations have been assumed as Dirichlet conditions on the River Cixerri, the lagoon, the sea, and the western side of the domain contour.

A_{15} The pumping effect has been considered as the Dirichlet condition on nodes of the two-dimensional mesh corresponding to given monitoring wells.

3.4   Mathematical Model and Its Numerical Solution

According to the hydrogeological and conceptual models, the mathematical model can be formulated as a coupled system of two partial differential equations, one describing mass conservation for the water-salt solution (flow equation), and the other mass conservation for the salt contaminant (the transport equation). Flow and transport equations are coupled by means of the constitutive equation relating density of the freshwater–saltwater mixture to salt concentration. Initial and Dirichlet, Neumann, or Cauchy boundary conditions are added to complete the mathematical formulation. The numerical solution code of these nonlinear equations involves spatial discretization with the finite element method following Galerkin's approach, and time discretization using finite differences; the nonlinear coupling is resolved using a Picard iteration method, which successively solves the flow and transport equations [Barrocu et al., 1994; Paniconi and Putti, 1995].

Results from the numerical code are expressed in terms of equivalent freshwater heads and normalized concentrations at selected time intervals and at each node of the three-dimensional mesh. The calibration-validation of the numerical solution code is performed by comparing the simulation results with hydrogeological and hydrogeochemical data obtained from the monitoring network.

Based on the geological features of the plain, four homogeneous zones have been defined. The first two are composed of the recent and ancient terraced alluvia of the Quaternary, the third of fractured granites and metamorphic schists of the Palaeozoic, and the fourth of fluvial and lacustrine sediments. Based on the aquifer hydrogeological parameters determined through pumping tests, hydraulic conductivity was set for each zone. On the other hand, it was necessary to reach an acceptable compromise between stability requirements of the code and soil properties reported in the literature so as to set longitudinal and transverse dispersivity values.

Moreover, the Dirichlet boundary conditions for flow and transport were defined on the basis of measured piezometric and salinity contour lines of January 1993, when the calibration-validation procedure started.

A two-dimensional mesh, based on the four zones identified in the domain and on the location of some significant wells within the monitoring network, was created with triangular elements. A three-dimensional mesh with tetrahedral elements is automatically generated by the numerical solution code.

Starting with uniform initial conditions and imposing boundary conditions, numerical tests were carried out until steady state conditions were reached. Heads and concentrations obtained at every node of the three-dimensional mesh were used as initial conditions for simulating the pumping effects. Pumping information on dug shallow and drilled wells was introduced as Dirichlet conditions. Then another steady state numerical test was carried out instead of the transient one, in view of the fact that the wells had been exploited for a long time so a dynamic equilibrium situation could actually be measured.

The water level contour lines for January 1993 could be fairly well reproduced with our calibration-validation procedure. Figure 7 shows the lateral recharge from the western boundary of the domain, the preferential flow direction toward the lagoon, the saltwork and the sea, and the zero contour line coinciding with the drawdown cones of some wells near the lagoon (phreatic aquifer) and the saltworks (confined aquifer).

The TDS contour lines plotted using the calculated data (Figure 8) and the field data for January 1993 show a similar trend. Water level and TDS contour lines, plotted using the field data measured in both the phreatic and confined aquifers, are shown in Figures 9 and 10 respectively.

At the end of the calibration-validation procedure, flow and transport simulations for March 1993, carried out with a numerical test for 90 days starting with January 1993, gave satisfactory results for both aquifers. In fact, the calculated contour lines in Figures 11 and 12 match fairly well the field contour lines shown in Figures 9 and 10.

4. ORGANIZATION OF THE GIS

4.1 Alphanumeric Database

The information collected in the monitoring network over the years of measurements was organized into an alphanumeric database, created with Microsoft Access software. To facilitate input, modification, and above all the display of the information contained in the database, masks were created for "coordinates and elevations," "water levels," and "chemical analyses" for

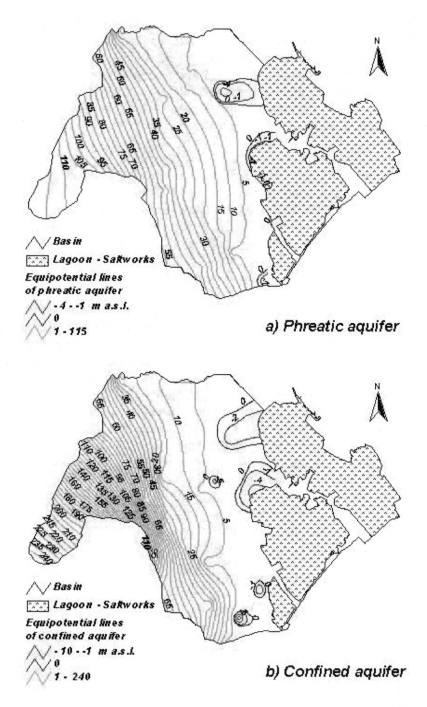

Figure 7: Calibration-validation of flow equation for January 1993.

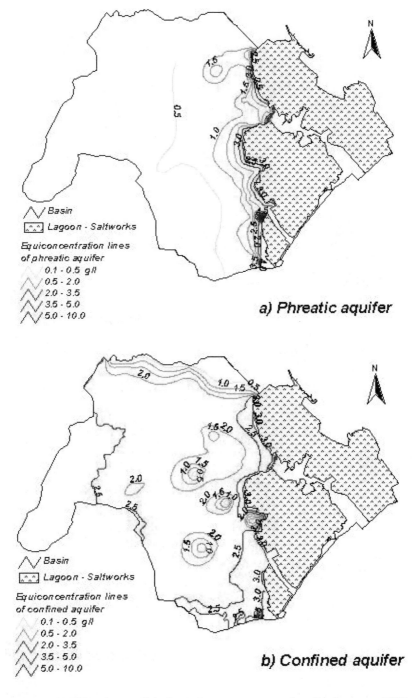

Figure 8: Calibration-validation of transport equation for January 1993.

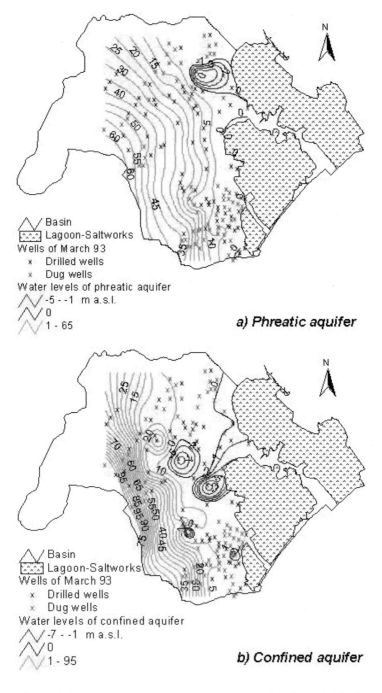

Figure 9: Water level contour lines measured in March 1993.

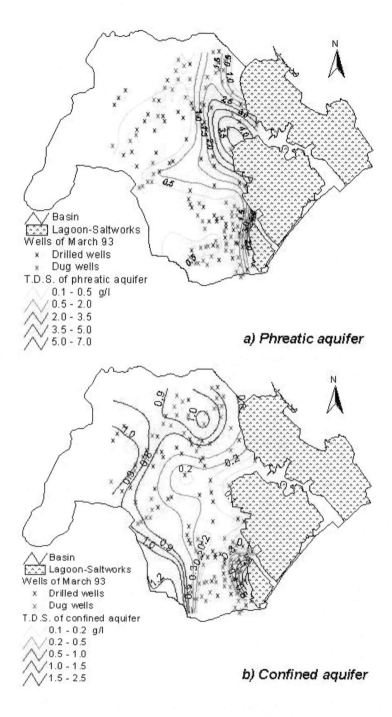

Figure 10: TDS contour lines measured in March 1993.

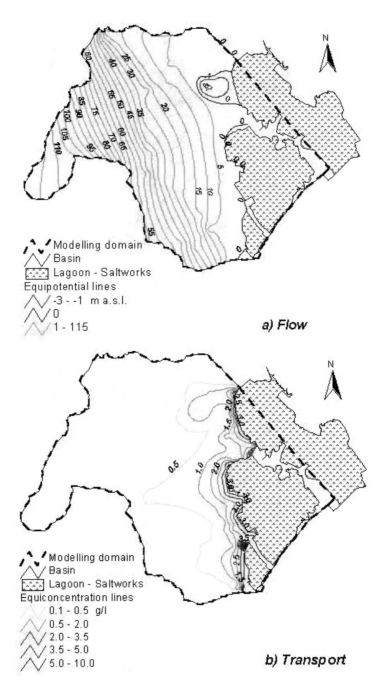

Figure 11: Flow and transport simulations of phreatic aquifer for
March 1993.

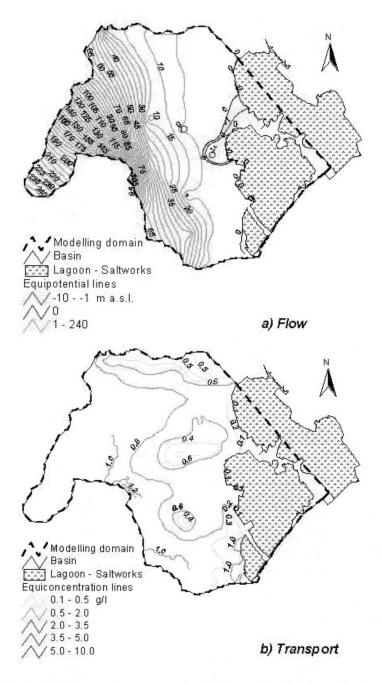

Figure 12: Flow and transport simulations of confined aquifer for
March 1993.

each month of measurement, and "pumping tests," linked together by action keys (see the accompanying CD). In each mask it is possible to search for information concerning a specific well, to access the other masks, to print the report of the data, after preview, and to return to the last mask displayed.

4.2   Geographic Database

The geographic database, created with ESRI ArcInfo and ArcView software, organizes all the information collected in the Capoterra alluvial plain into a number of views and themes (see the accompanying CD):

- a view that contains levels of general information that were created from .DXF files in AutoCad and imported and converted into ArcInfo coverage and ArcView shape, i.e., geology, elevation, surface hydrography, monitoring network;
- as many views, constructed querying the Access database by using an SQL connection feature, as there are months of measurement, containing the levels of information on water level, TDS, and electric conductivity contour lines of the phreatic and confined aquifers, and the spatial distribution of the principal physico-chemical parameters, such as temperature, pH, cations and anions;
- a view containing the levels of information created from the saltwater intrusion modeling results including the two-dimensional mesh, the equipotential and equiconcentration lines of the aquifers both for the calibration-validation procedure and the simulations.

5.   CONCLUSIONS

With regard to the application of the modeling procedure in the alluvial plain of Capoterra, the hydrogeological and conceptual models have been fairly well defined, and both the mathematical model and numerical solution are reliable.

Implementation of this modeling procedure in such a complex physical and anthropogenic context presupposes a sound knowledge base composed of the data collected in the monitoring network purposely set up in the area.

The numerical solution of the mathematical model has demonstrated that the computational capacity of the code can be fully exploited if the possibility of using a point data base exists both for describing the geometry of the domain identified and characterizing the process in physical terms.

The results obtained have pointed to the need for integrating modeling with GIS systems so as to facilitate collection, updating, and interpretation of the data. GIS also aids model construction and implementation and the critical evaluation of the results for achieving an efficient decision support system.

REFERENCES

Barrocu, G., Fidelibus, M.D., Sciabica, M.G. and Uras, G., "Hydrogeological and hydrogeochemical study of saltwater intrusion in the Capoterra coastal aquifer system (Sardinia)," Atti del 13th SWIM (Salt Water Intrusion Meeting), Villasimius-Cagliari, ed. G. Barrocu, 105–111, Giugno, 1994.

Barrocu, G., Sciabica, M.G. and Paniconi, C., "Three-Dimensional Model of Saltwater Intrusion in the Capoterra Coastal Aquifer System (Sardinia)," Atti del 13th SWIM (Salt Water Intrusion Meeting), Villasimius-Cagliari, ed. G. Barrocu, 77–86, Giugno, 1994.

Barrocu, G., Muscas, L. and Sciabica, M.G., "Prima esperienza di sviluppo di un Sistema Informativo Geografico della Piana di Capoterra (Sardegna)," In: Proc. 4th Nat. Conf. della Federazione A.S.I.T.A. (Associazioni Scientifiche per le Informazioni Territoriali e Ambientali), 1.423–1.428, Genova, 2000.

Barrocu, G., Sciabica, M.G., Uras, G., Cortis, A. and Vernier, E., "Saltwater intrusion and artificial recharge modeling in the coastal aquifer system of Capoterrra (Southern Sardinia)." In: Proc. Int. Conf. on Water Problems in the Mediterranean Countries, Near East University Civil Engineering Department Nicosia, 2.1001–2.1007, North Cyprus, 1997.

Barrocu, G., Sciabica, M.G. and Uras, G., "Transport modeling of saltwater intrusion processes in the coastal aquifer system of Capoterra (Southern Sardinia, Italy)." In: Proc. 15th SWIM (Salt Water Intrusion Meeting), 23–27, Ghent, Belgio, 1998.

Barrocu, G., Muscas, L. and Sciabica, M.G., "GIS and Modeling Finalized to Studying Saltwater Intrusion in the Capoterra Alluvial Plain (Sardinia-Italy)," First International Conference on Saltwater Intrusion and Coastal Aquifers—Monitoring, Modeling, and Management, Essaouira, Morocco, eds. D. Ouazar and A.H.-D. Cheng, April 23–25, 2001.

Bear, J. and Bachmat, Y., *Introduction to Modeling of Transport Phenomena in Porous Media*. Kluwer Academic, Dordrecht, Holland, 1990.

Issar, A. and Passchier, R., "Regional Hydrogeological Concepts. Part II, Dispense dei Corsi in Hydrogeological Modeling of Flow and Pollution in Dry Regions," Jacob Blaustein Institute for Desert Research, Ben-Gurion University of the Negev, Israel, 1992.

Paniconi, C. and Putti, M., "Picard and Newton linearization for the couplet model of saltwater intrusion aquifers," *Adv. Water. Resour.*, **18**(3), 159–170, 1995.

Sciabica, M.G., "Validazione dei modelli concettuali e matematici degli acquiferi con i dati idrogeologici ed idrogeochimici delle reti di monitoraggio," Ph.D. thesis, Politecnico di Torino and University of Cagliari, 1994.

Vernier, E., "Studio idrogeologico ed idrogeochimico del sistema acquifero della piana alluvionale di Capoterra (Sardegna meridionale)," Ph.D. thesis, University of Cagliari, 1999.

CHAPTER 10

Uncertainty Analysis of Seawater Intrusion and Implications for Radionuclide Transport at Amchitka Island's Underground Nuclear Tests

A. Hassan, J. Chapman, K. Pohlmann

1. INTRODUCTION

All studies of subsurface processes face the challenge presented by limited observations of the environment of interest. By its very nature, the detailed characteristics of the subsurface are hidden and data collection efforts are generally hindered by technical and financial constraints. The result is that uncertainty is a factor in all groundwater studies. Seawater intrusion environments present both special opportunities and special challenges for incorporating uncertainty into numerical simulations of groundwater flow and contaminant transport. Opportunities come from the constraints that the seawater–freshwater system provides; challenges come from the numerically intensive solutions demanded by simultaneous solution of the energy and mass transport equations.

The impact of uncertainty in the analysis of contaminant transport in coastal aquifers is an important aspect of evaluating radionuclide transport from three underground nuclear tests conducted by the U.S. on Amchitka Island, Alaska. Testing was conducted in the 1960s and very early 1970s on the Aleutian island to characterize the seismic signals from underground tests in active tectonic regimes, and to avoid proximity to high-rise buildings and resulting ground motion problems. As the U.S. Department of Energy focused on environmental management of nuclear sites in the 1990s, a decision was made to revisit contaminant transport predictions for the island, taking advantage of the advances in the understanding of island hydraulic systems and in computational power that occurred in the decades after the tests.

Though the general geologic conditions are similar for the three tests, they differ in their depth and thus position relative to the freshwater–seawater transition zone (TZ). Amchitka is a long, thin island separating the Bering Sea and Pacific Ocean, predominantly consisting of Tertiary-age

1-56670-605-X/04/$0.00+$1.50
© 2004 by CRC Press LLC

submarine and subaerially deposited volcanic rocks. The tests all occurred in the lowland plateau region of the island, with the lithologic sequence dominated by interbedded basalts and breccias. The shallowest test is Long Shot, conducted in 1965 at a depth of 700 m. The Milrow test occurred next, in 1969, at a depth of 1,220 m. The deepest test was Cannikin at 1,790 m, conducted in 1971.

There are strongly developed joint and fault systems on Amchitka and groundwater is believed to move predominantly by fracture flow between matrix blocks of relatively high porosity. The subsurface is saturated to within a couple of meters of ground surface, and the lowland plateau has many lakes, ponds, and streams. Hydraulic head decreases with increasing depth through the freshwater lens, supporting the basic conceptualization of freshwater recharge across the island surface with downward-directed gradients to the transition with seawater. Samples of groundwater from exploratory boreholes at each site indicate that Long Shot was detonated in the freshwater lens and Milrow was below the TZ. The data from Cannikin are equivocal, and though Cannikin is deeper than Milrow, the possibility of asymmetry in the freshwater lens precludes extrapolation. In addition to chemical data from wells and boreholes, numerous packer tests were performed and provide hydraulic data, and abundant cores were collected and analyzed for transport properties (such as porosity and sorption).

The conceptual model of flow for each site is governed by the principles of island hydraulics. Recharge of precipitation on the ground surface maintains a freshwater lens by active circulation downward and outward to discharge on the sea floor. Below the TZ, salt dispersed into the TZ and discharged from the system is replaced by a very low velocity counter-circulation, recharged by infiltration along the sea floor far beyond the beach margin, past the freshwater discharge zone. A groundwater divide is assumed to exist, coincident with the topographic divide, separating flow to the Bering Sea (applicable for Long Shot and Cannikin) from flow to the Pacific Ocean (Milrow). The simplicity of the island hydraulic model is enhanced by the absence of pumping or any form of groundwater development on the island, so that steady-state conditions are assumed. Figure 1 shows a map of Amchitka Island and the location and perspective of each of the three cross sections representing the simulation domains for the three tests.

2. PROCESSES MODELED, PARAMETERS, AND CALIBRATION

Modeling Amchitka's nuclear tests encompasses two major processes: 1) the flow modeling, taken here to include density-driven flow,

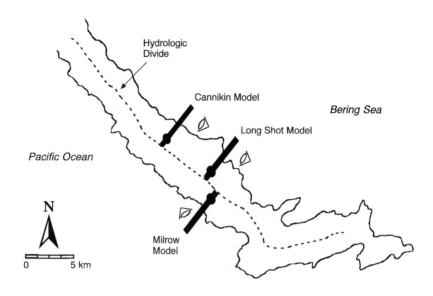

Figure 1: Location of model cross section for each site with the cartoon eye
indicating the perspective of subsequent figures.

saltwater intrusion, and heat-driven flow, and 2) the contaminant transport
modeling, combining radioactive source evaluation and decay, retardation
processes, release functions, and matrix diffusion. The symmetry of island
hydraulics lends itself to considering flow in two dimensions, on a transect
from the hydrologic divide along the island's centerline, through the nuclear
test location, and on to the sea. The boundary conditions for the flow
problem entail no flow coinciding with the groundwater divide and along the
bottom boundary. The seaward boundary is defined by specified head and
constant concentration equivalent to seawater. The top boundary has two
segments. The portion across the island receives a recharge flux at a
freshwater concentration, and the portion along the ocean is a specified head
dependent on the bathymetry. Figure 2 shows the Milrow topographic and
bathymetric profile, the domain geometry and boundary conditions, and the
finite element mesh used to discretize the density-driven flow equations. The
mesh is refined in the entire left upper triangle of the simulation domain
since the TZ varies widely with the random parameters selected.

    For the other two sites, similar domain geometry and boundary
conditions are utilized. However, the upper boundary is determined based on
the specific site's topography and bathymetry, which is slightly different

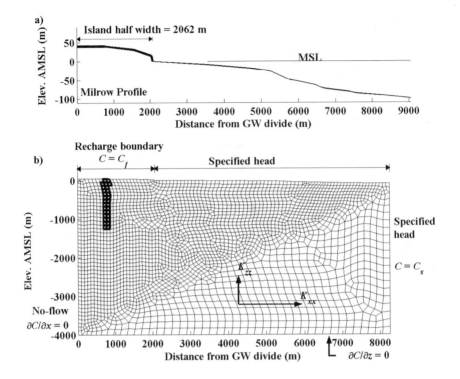

Figure 2: Milrow profile that determines (a) the upper boundary of the simulation domain, and (b) the discretization and boundary conditions.

among the three sites. Island-specific data are used to constrain the parameter values used to construct the seawater intrusion flow problem. Hydraulic conductivity, K, data collected from six boreholes are used to yield the best estimate for a homogeneous conductivity value and the range of uncertainty associated with this estimate. The geologic environment suggests strong anisotropy, so that vertical hydraulic conductivity, K_{zz}, is assumed to be one-tenth the horizontal value (except in the chimney above the nuclear cavity, where collapse is assumed to increase K_{zz} relative to that of the horizontal conductivity, K_{xx}). Temperature logs measured in several boreholes and water balance estimates are used to derive groundwater recharge, R, values. Measurements of total porosity on almost 200 core samples from four boreholes provided a mean and distribution for matrix porosity. No measurements of fracture porosity, a notoriously difficult-to-measure parameter, are available, so literature values guided that selection. The transport model also required data on retardation properties, which were obtained using sorption and diffusion experiments from core material.

The model for each of the nuclear tests was calibrated using site-specific hydraulic head and water chemistry data. The objective of the calibration was to select base-case, uniform flow, and saltwater intrusion parameters that yield a modeling result as close as possible to that observed in the natural system. Difficulty was encountered in obtaining simultaneous best-fits to the two targets (head and chemistry). The best parameters to match head would result in a less perfect match for the chemical profile, and vice versa. The critical calibration feature for locating the mid-point of the TZ is the ratio of recharge to hydraulic conductivity (R/K). Macrodispersivity controlled the width of the TZ modeled around the mid-point. Ultimately, compromises were made to achieve the optimum fit to both heads and chemistry, and more weight was given to the hydraulic head measurements due to reported difficulties encountered in obtaining representative samples from these very deep boreholes during drilling operations. The configuration of the seawater interface differs from one site model to another, with a deeper freshwater lens calculated on the Bering Sea side of the island.

3.   PARAMETRIC UNCERTAINTY ANALYSIS

To optimize the modeling process, a parametric uncertainty analysis was performed to identify which parameters are important to treat as uncertain in the flow and transport modeling and which to set as constant, best estimate, values. This analysis was performed for the Milrow site and the findings are applied to all sites. The processes evaluated through their flow and transport parameters include recharge, saltwater intrusion, radionuclide transport, glass dissolution, and matrix diffusion. The end result of this analysis is a relative comparison of the effect of uncertainty of each individual parameter on the final transport results in terms of the arrival time and mass flux of radionuclides crossing the seafloor.

3.1   Uncertainty Analysis of Flow Parameters

The parameters of concern here are the hydraulic conductivity, K, the recharge, R, and the longitudinal and transverse macrodispersivities, A_L and A_T. Since the saltwater intrusion problem encounters a density-driven flow, the macrodispersivities are considered as flow parameters. In addition, the porosity is also considered at this stage as the spatial variability of porosity between the chimney and the surrounding area affects the solution of the saltwater intrusion problem. In all cases, the flow and the advection-dispersion equations are solved simultaneously until a steady-state condition is reached. The solution provides the groundwater velocities and the concentration distribution that can be used to identify the location and

First Modeling Stage (Parametric Uncertainty Analysis)

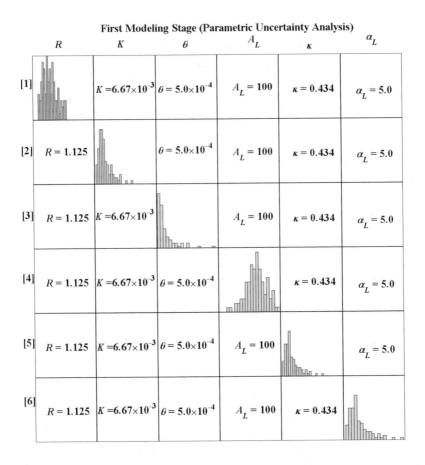

	R	K	θ	A_L	κ	α_L
[1]		$K = 6.67 \times 10^{-3}$	$\theta = 5.0 \times 10^{-4}$	$A_L = 100$	$\kappa = 0.434$	$\alpha_L = 5.0$
[2]	$R = 1.125$		$\theta = 5.0 \times 10^{-4}$	$A_L = 100$	$\kappa = 0.434$	$\alpha_L = 5.0$
[3]	$R = 1.125$	$K = 6.67 \times 10^{-3}$		$A_L = 100$	$\kappa = 0.434$	$\alpha_L = 5.0$
[4]	$R = 1.125$	$K = 6.67 \times 10^{-3}$	$\theta = 5.0 \times 10^{-4}$		$\kappa = 0.434$	$\alpha_L = 5.0$
[5]	$R = 1.125$	$K = 6.67 \times 10^{-3}$	$\theta = 5.0 \times 10^{-4}$	$A_L = 100$		$\alpha_L = 5.0$
[6]	$R = 1.125$	$K = 6.67 \times 10^{-3}$	$\theta = 5.0 \times 10^{-4}$	$A_L = 100$	$\kappa = 0.434$	

Second Modeling Stage (Combined Parametric Uncertainty)

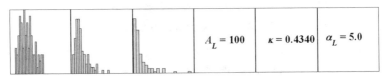

R	K	θ	A_L	κ	α_L
			$A_L = 100$	$\kappa = 0.4340$	$\alpha_L = 5.0$

Figure 3: A summary of the two modeling stages and the implementation of the parametric uncertainty analysis. The numbers in square brackets are for the scenarios studied in the first modeling stage.

thickness of the TZ. For each of the four parameters, a random distribution of 100 values below and above a "mean" value close to the calibration result is generated. Figure 3 summarizes the parametric uncertainty analysis for all parameters (first modeling stage) and the combined uncertainty analysis (second modeling stage).

For the first modeling stage, a lognormal distribution was used to generate the recharge values for Scenario 1 and the distribution was truncated such that the upper and lower limits lead to reasonable TZ movement around the location indicated by the chemistry data. For Scenario 2, the uncertain conductivity values are generated from a lognormal distribution and have a mean value of 6.773×10^{-3} m/day, which is equivalent to the Milrow calibration value. As for recharge, a lognormal distribution was selected with upper and lower limits that were consistent with the data and yielded a reasonable TZ.

From these conductivity limits and those of the recharge, the recharge-conductivity ratio is changing from 1.35×10^{-3} to 9.05×10^{-3} for Scenario 1 and from 1.26×10^{-3} to 2.05×10^{-2} for Scenario 2. It should be mentioned here that the recharge-conductivity ratio is the factor that controls the location of the TZ, but the magnitude of the velocity depends on the recharge and conductivity values. The large macrodispersivity values are considered to account for the additional mixing resulting from spatial variability that is not considered in the model and to avoid violation of the Peclet number if small macrodispersivity values are used. For all cases considered, the chimney and cavity porosity is set to a fixed value of 0.07, whereas the rest of the domain is assigned a fracture porosity value that is obtained from a random distribution having a minimum value of 1.294×10^{-5} and a maximum value of 3.8×10^{-3}.

Having generated the individual random distributions for each of the parameters considered, the variable-fluid-density groundwater flow problem is solved using the FEFLOW code [Diersch, 1998]. For each one of the four parameters considered, a set of 100 steady-state velocity and concentration distributions is obtained that corresponds to the 100 random input values. For the simulated head and concentration values at the Milrow calibration well, Uae-2, the mean of the 100 realizations as well as the standard deviation of the result are computed.

Figures 4 and 5 show the impact of the extreme values of R and K on the TZ location for Scenarios 1 and 2 that address the uncertainty in R and K, respectively. The smaller range of R/K is reflected on the TZ locations shown in Figure 4. Figures 6 and 7 show the sensitivity of the concentration and head to the uncertainty in the values of recharge and conductivity, respectively. In each figure, the mean of the Monte Carlo runs, the mean ± one standard deviation, and the data points are plotted. It can be seen that for the recharge case, the one standard deviation confidence interval around the mean captures most of the data points for concentration and for head

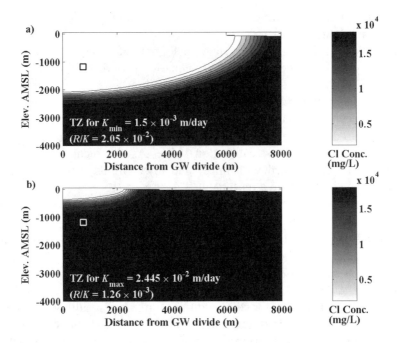

Figure 4: Transition zone location relative to cavity location for the
extreme values of R in the recharge sensitivity case.

measurements. The conductivity case (Figure 7) covers the high
concentration data (saltwater side) but gives lower concentrations than the
data for the freshwater side of the TZ. The head sensitivity to conductivity
variability shown in Figure 7 indicates that the confidence interval
encompasses all the head data at Uae-2.

The porosity does not affect the solution of the flow problem even
with the chimney having a different porosity. The porosity only influences
the speed at which the system converges to steady state, and as such,
simulated heads and concentrations at Uae-2 do not show any sensitivity to
the fracture porosity value outside the chimney. It should be recognized,
however, that the fracture porosity outside the chimney and cavity area will
have a dramatic effect on travel times and radioactive decay of mass released
from the cavity and migrating toward the seafloor. The range of 60 to 500 m
considered for A_L has a minor effect on the head and concentration at Uae-2,
especially at the center of the TZ.

Again, the final decision as to whether the uncertainty in a parameter
is important to include in the final modeling stage cannot be determined from
these results. The criterion for selecting the most influential parameters can

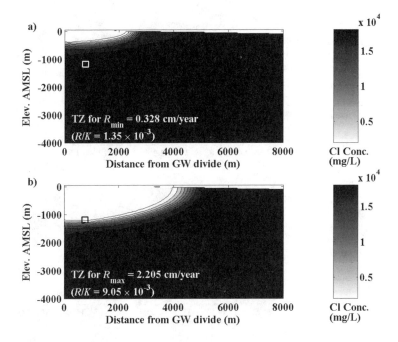

Figure 5: Transition zone location relative to the cavity location for the extreme values of K in the conductivity sensitivity case.

only be determined by analyzing the transport results in terms of travel times from the cavity to the seafloor and location where breakthrough occurs. The set of results discussed here indicates that the simulated heads and concentrations at Uae-2 are most sensitive to conductivity and recharge and least sensitive to fracture porosity outside the chimney and macrodispersivity. The parameter importance to the transport results may be confirmed or changed by analyzing the travel time statistics for particles originating from the cavity and breaking through the seafloor.

The velocity realizations resulting from the solution of the flow problem are used to model the radionuclide transport from the cavity toward the seafloor. The transport parameters are kept fixed at their means while addressing the effect of the four parameters that change the flow regime. When the effect of transport parameters, such as matrix diffusion coefficient, glass dissolution rate, etc., is studied, a single velocity realization with the flow parameters fixed at the calibration values is used.

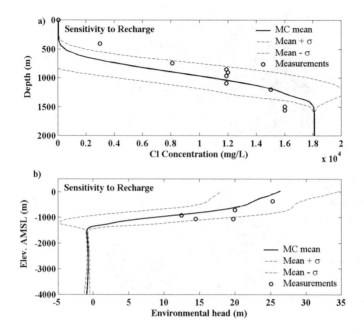

Figure 6: Sensitivity of modeled concentrations and heads at Uae-2 to the recharge uncertainty.

3.2   Uncertainty Analysis of Transport Parameters

To analyze the effect of transport parameters' uncertainty on transport results, a 100-value random distribution for local dispersivity, α_L, is generated from a lognormal distribution. The analysis is performed using a single flow realization and the transport simulations are performed for 100 different α_L values. A similar analysis is performed to analyze the effect of the matrix diffusion parameter, κ. Based on available data and literature values, a best estimate for κ of 1.37 day$^{-1/2}$ was derived. This value leads to a very strong diffusion into the matrix, which significantly delays the mass arrival to the seafloor, producing no mass breakthrough at the seafloor within the selected time frame of about 27,400 years of this first modeling stage. As there is a large degree of uncertainty in determining this parameter and the uncertainty derived by the conceptual model assumptions for diffusion (e.g., assumption of an infinite matrix), values for κ that are smaller than the best estimate of 1.37 were chosen. A random distribution of 100 values is generated for κ with a minimum of 0.0394, a maximum of 1.372, and a mean of 0.352.

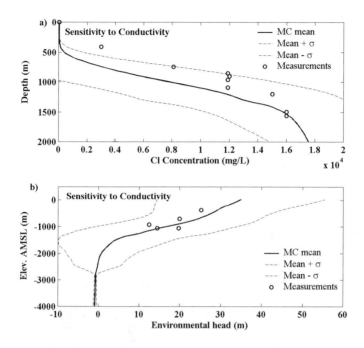

Figure 7: Sensitivity of modeled concentrations and heads at Uae-2 to the conductivity uncertainty.

The transport simulations are performed using a standard random walk particle tracking method [Tompson and Gelhar, 1990; Tompson, 1993; LaBolle *et al.*, 1996, 2000; Hassan *et al.*, 1997, 1998]. For more details about the transport simulations that are pertinent to this study, the reader is referred to Hassan *et al.* [2001] and Pohlmann *et al.* [2002].

To show how the particles travel from the cavity to the seafloor (breakthrough plane), a single realization showing about 100% mass breakthrough during a time frame of 2,200 years is selected for analysis and visualization. The particle locations at different times are reported and used to visualize the plume shape and movement. Figure 8 shows three snapshots of the particles' distribution at different times with the percentage mass reaching the seafloor computed and presented on the figure. No particles reach the seafloor within the first 100 years after the detonation. At 140 years, the leading edge of the plume starts to arrive at the seafloor. Larger numbers of particles arrive between 140 and 180 years, with a total of 1.2% of the initial mass reaching the seafloor by 180 years. For the rest of

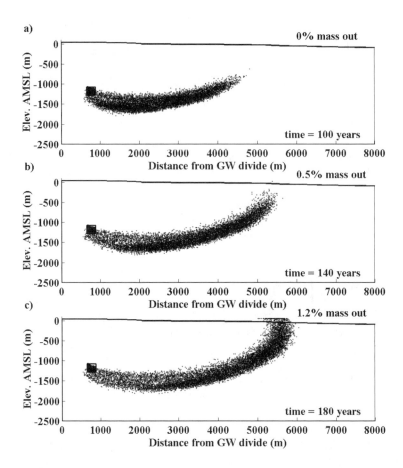

Figure 8: Snapshots of the particles' locations showing how the plume moves along the TZ of the seawater intrusion problem.

simulation time, the accompanying CD for this book contains an animated movie showing the plume movement as a function of time.

3.3   Results of the Parametric Uncertainty Analysis

The mass flux breakthrough curves resulting from the arrival of radionuclides to the seafloor are analyzed in terms of the mean arrival time of the mass that breaks through within the simulation time frame and the location of this breakthrough along the bathymetric profile. Recall that the purpose of this analysis is to select the parameters for which the associated uncertainty has the most significant effect on transport results expressed in terms of uncertainty of travel time to the seafloor and the location where breakthrough occurs. By doing so, the parameters for which the uncertainty only slightly affects the uncertainty in travel time and transverse location of the breakthrough can be identified, and as such, these parameters are fixed at their best estimate and only those with significant effects are varied.

The results of the sensitivity analysis performed for the six parameters, K, R, θ, A_L (in saltwater intrusion), a_L (in radionuclide transport modeling), and κ are summarized in Table 1, in which the matrix diffusion parameter, κ, is assigned the value of 0.0434 day$^{-1/2}$ that is one order of magnitude lower than the base-case value. This is because the base-case value leads to significant matrix diffusion that conceals the different uncertainty effects. Reducing the effect of matrix diffusion allows for a comparison between different uncertainties and their impact on the transport results. For each case, the table presents the range of values of the input parameter (minimum, maximum, and mean), the standard deviation, and the coefficient of variation. On the output side, the results are presented in terms of the statistics of travel time and transverse location where breakthrough occurs. For each single realization of the radionuclide transport, the mean arrival time and mean transverse location of the mass that has crossed the seafloor within 27,400 years are recorded. The resulting ensemble of these values is used to compute the mean, standard deviation, and coefficient of variation of the travel time and location, which are presented in Table 1.

To facilitate the comparison between different cases, one would compare the values of the coefficient of variation on both input and output sides. Among the six cases in Table 1, the two cases encountering variability in the macro/local dispersivity value lead to very small uncertainty in the travel time and the transverse location in comparison to other parameters. Although the coefficient of variation of a_L in radionuclide transport simulations is higher than that of conductivity and recharge, the resulting coefficients of variation for travel time and transverse location are much smaller. Therefore, it can be argued that the uncertainty in these two parameters may be neglected, as their variabilities slightly influence transport results when compared to other parameters. This leaves the four parameters, K, R, θ, and κ. The fracture porosity variability with the highest coefficient of variation among these four parameters leads to the highest variability in mean arrival time. The conductivity, on the other hand, leads to the highest variability in transverse location. The first three parameters of this reduced list influence the solution of the flow problem and thus require multiple realizations of the flow field. The matrix diffusion parameter is a transport parameter that does not require multiple flow realizations.

		Parameters					
		K (m/d)	R (cm/y)	A_L (m)	θ (-)	α_L (m)	κ (d$^{-1/2}$)
Input Statistics	Min	0.89×10^{-3}	0.328	62	1.3×10^{-3}	0.56	0.039
	Mean	6.77×10^{-3}	1.125	300	5.2×10^{-3}	5.0	0.352
	Max	2.45×10^{-2}	2.205	500	3.8×10^{-3}	19.5	1.37
	σ	4.34×10^{-3}	0.475	82	6.4×10^{-4}	3.45	0.243
	cv	0.641	0.422	0.27	1.23	0.69	0.691
Travel Time (10^3 years)	Mean	22.19	22.00	20.65	19.101	23.0	25.77
	σ	1.98	3.484	0.742	4.965	0.31	1.15
	cv	0.089	0.158	0.036	0.260	0.01	0.045
BT Location (km)	Mean	3.629	3.404	3.394	3.382	3.37	3.274
	σ	0.660	0.375	0.009	0.042	0.02	0.031
	cv	0.182	0.110	0.003	0.012	0.01	0.009

Table 1: Results of the uncertainty analysis comparing the effects of different parameters on plume travel time and transverse location of the breakthrough.

The final choice for the uncertain parameters for the second modeling stage is the three flow parameters. This choice is motivated by the fact that the available data only pertain to the solution of the flow problem and can be used to guide the generation of the random distributions in the second stage. Head and chloride concentration data can be used as criteria for determining whether the combined random distributions lead to realistic flow solutions or not. Given that using the same random distribution for κ as in the first stage or skewing it toward higher or lower values cannot be judged or tested against data, the transport results are obtained using a conservative estimate for the κ, which is kept constant in all subsequent analysis.

3.4   Flow and Transport Results of the Second Modeling Stage

For the primary flow and transport modeling for the sites, using the significant uncertain parameters identified in the parametric uncertainty analysis (K, R, and θ), the same model meshes employed in the individual parametric uncertainty analysis are used. Three new random distributions are generated for the conductivity, recharge, and fracture porosity for each site with the total number of realizations between 240 and 300. Flow and transport simulations are performed in a manner similar to the first stage. The output presented here is a point mass flux distribution as a function of space and time, $q(x, t)$, where q is the mass crossing a unit cross-sectional area per unit time, x is the horizontal distance along the seafloor relative to the island center (or groundwater divide), and t is the time since the migration started. This two-dimensional distribution of q is obtained for each individual realization and the ensemble mean, $<q>$, is obtained by averaging over all realizations for each site.

Figure 9 shows the transport results for the three sites with Milrow in the top plot, Cannikin in the middle plot, and Long Shot in the lower plot. In each plot, the cavity location (source of radionuclides) and the TZ locations associated with the extreme values of R/K among all realizations are shown. The plots also show the space-time distribution of the ensemble mean of the point mass flux, $< q(x, t) >$, for carbon-14 (half life = 5,730 years) with the right axis indicating time. This plot is superimposed on the TZ plots to show the location where breakthrough occurs relative to the cavity location and the limits of the TZ location produced by the uncertain input parameters.

The figure shows that the incorporation of uncertainty in the TZ location (through uncertainty in recharge and hydraulic conductivity), combined with the different location of the test cavity between the three sites, leads to a large variation in transport results from one test to the other. The transport results calculated for a realization with the cavity intersecting the TZ is dramatically different than for a realization with the TZ below the cavity. For both Milrow and Cannikin, the early-time portion of the mass flux breakthrough is dominated by the realizations representing the transition zone at or below the cavities. Based on the results shown in Figure 10, the Long Shot cavity is always located at the freshwater side and very far from the center of the transition zone. This leads to the direct movement of radionuclides from the cavity toward the seafloor. The Milrow cavity and that of Cannikin, on the other hand, are located at the saltwater side of the TZ in many realizations. This means that in these realizations, the cavity comes in contact with the very slow flow pattern occurring at the lower edge of the TZ. This explains why a number of realizations at Milrow and Cannikin do not produce any mass breakthrough within 2,200 years. For Cannikin, the cavity is deeper than that for Milrow. This results in a longer flow path to the seafloor, thereby causing breakthrough to occur at a later time and with smaller mass flux values than Milrow due to the increased radioactive decay. The location of the breakthrough is mainly dominated by the cavity location; thus it can be seen that the breakthrough at Long Shot is closest to the shoreline followed by that of Milrow and then Cannikin.

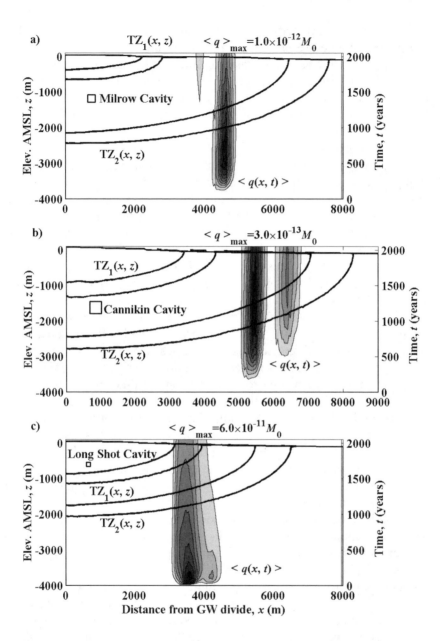

Figure 9: Expected value of point mass flux, $< q(x, t) >$, at the three sites as a function of breakthrough time (right vertical axis) and location (horizontal axis). The TZ location (left vertical axis and horizontal axis) for maximum and minimum values of R/K is shown to relate to the cavity location and the breakthrough location.

Parameter	Value
Rock Volumetric Heat Capacity, $\rho_s c_s$	1.9×10^6 J/m^{3}C
Water Volumetric Heat Capacity, $\rho_0 c_0$	4.2 J/m^{3}C
Rock Thermal Conductivity, λ_s	2.59 J/m^{3}C
Water Thermal Conductivity, λ_0	0.56 J/m^{3}C
Thermal Longitudinal Dispersivity, β_L	100 m
Thermal Transverse Dispersivity, β_T	10 m
Water Density and Viscosity, ρ_0 and μ_0	6th order function of temperature

Table 2: Values of parameters used in FEFLOW for simulations incorporating geothermal heat.

4. SENSITIVITY STUDIES

Numerical modeling of the coastal aquifer systems at Amchitka Island directly incorporates uncertainties in critical parameters where data allow. However, some uncertainties cannot be addressed through that process, either due to lack of data, or because the uncertainty is in the underlying conceptual model or numerical approach. These uncertainties are addressed through separate sensitivity studies and are discussed in the following sections.

4.1 Geothermal Heat

The base-case flow models are run under isothermal conditions, assuming that compared to geothermal effects, the freshwater–seawater dynamics dominate the island flow system. The impacts of including geothermal heat are addressed through a nonisothermal analysis of the Milrow flow system, where hydraulic head, concentration, and temperature data sets are most complete and reliable.

The geothermal model simulates pre-nuclear test conditions; therefore, the chimney is not included and K and θ are treated as homogeneous properties throughout the domain. With the exceptions noted below, values of the groundwater flow parameters are the same as the values used in the calibrated flow model of Milrow. The values of the parameters required for the geothermal component are listed in Table 2. Fluid density and viscosity are dependent on both concentration and temperature, based on a nonlinear relationship of density to temperature incorporated in the FEFLOW code. Rock thermal properties are based on core samples from the island [Green, 1965]. The thermal properties of water are FEFLOW default values. The temperature of 125°C at the bottom boundary is extrapolated

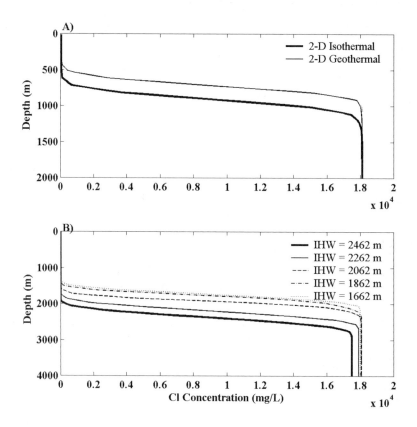

Figure 10: Effect of the inclusion of geothermal heat (A) and island half
width, IHW (B) on the two-dimensional TZ at Milrow.

from temperature profiles measured in several Amchitka boreholes [Sass and
Moses, 1969], and indicates a geothermal gradient of 3.2°C per 100 m depth.
The temperature at the upper boundary is 4°C, which is consistent with both
the mean average air temperature noted for Amchitka [Armstrong, 1977] and
the value for ground surface extrapolated from the subsurface temperature
profiles.

   The results indicate that thermally driven buoyant flow caused by the
geothermal gradient increases the vertical upward flux below the island and
shifts the transition zone almost 200 m higher relative to the isothermal case
(Figure 10A). At the TZ, this increased vertical flux is then directed seaward,
resulting in higher velocities along the TZ as compared to the isothermal
case. Despite these differences, the overall patterns of flow are similar to the
isothermal case. The upward and left (toward the divide) components of

velocities simulated below the TZ are both larger due to the buoyancy-driven flow simulated in the geothermal model. Higher flow rates mean that velocities near the working point, which is located below the TZ at Milrow, are higher when including the effects of geothermal heat. The vertical and horizontal velocities at the Milrow working point are about twofold higher in the geothermal model. Velocities higher than the isothermal model are generally maintained along the predicted flowpaths from the working point toward the sea, suggesting that inclusion of geothermal heat in the model simulations has the effect of reducing contaminant travel times for the Milrow and Cannikin sites where the working points are below the TZ in many of the realizations considered.

4.2   Island Half-Width

The conceptual model for groundwater flow at Amchitka assumes that a groundwater divide runs along the long axis of the island, separating flow to the Bering Sea on one side and flow to the Pacific Ocean on the other (see Figure 1). The position of the divide is also assumed to coincide with that of the surface water divide. This assumption can be called into question due to the observation of asymmetry in the freshwater lens beneath the island [Fenske, 1972a, b]. This asymmetry is supported by the data analysis and modeling performed here, which suggests that the freshwater lens is deeper at Long Shot and Cannikin than at Milrow.

Not only is there uncertainty as to whether the groundwater and surface water divides coincide, there is additional uncertainty in the location of the surface water divide itself, as the topography of the island in the area of the nuclear tests is very subdued. The surface water divide was estimated using a detailed series of topographic maps at a scale 1:6,000 and with a 10-foot contour interval. Despite this resolution, the distance between 10-foot elevation contours can reach over 100 m in places.

To understand the impact of this uncertainty on the groundwater modeling, several sensitivity cases were evaluated. In these, the island half-width was assumed to be 200 and 400 m wider than the estimate for Milrow, and also assumed to be 200 and 400 m narrower than used in the base-case model. For reference, the base-case half-width used at Milrow is 2,062 m, so that plus and minus 10 and 20% differences are considered here. One realization was used for these calculations, one in which the cavity is located in the freshwater lens. It shows a 100% mass breakthrough and has the parameter values $K = 2.34 \times 10^{-2}$ m/d, $R = 1.82$ cm/yr, and $\theta = 1.62 \times 10^{-4}$.

Varying the island half-width both affects the depth to the TZ (through varying the land surface available for recharge) and the position of the cavity in the flow system (by virtue of changing the distance from the test to the no-flow boundary). The TZ depicted from the vertical chloride concentrations in the Uae-2 well at Milrow is plotted in Figure 10B for the base-case island width and the four additional sensitivity cases. Reducing the island half-width decreases the depth of the TZ, and cuts the distance between the cavity and the transition in half for the 400-m-shorter half-width. Conversely, the TZ is deepened by an increasing half-width, increasing the distance from the cavity to the TZ by a factor of two for the 400-m-wide island. The flowpath distance to the seafloor from the cavity is also affected, lengthening for a wider island and shrinking for a smaller one.

The impact of these various configurations on transport is also investigated. It is found that the 400-m-longer half-width leads to an earlier breakthrough of mass at a peak flux about two times larger than the base case. On the other hand, the 400-m-shorter half-width results in a delay in breakthrough at a peak mass about five times lower than the base case.

4.3   Dimensionality of Rubble Chimney

The models used in the uncertainty analysis utilize a two-dimensional perspective to analyze the flow and transport problem, a simplification that is consistent with the island hydraulic environment. This simplifying assumption is considered reasonable for the conceptual model and is significantly more computationally efficient than a fully three-dimensional formulation. However, the two-dimensional formulation accounts for the geometry of the rubble chimney only in the plane of the model, i.e., parallel to the natural flow direction, and therefore the chimney is simulated as extending infinitely in the direction perpendicular to the plane of the model. In reality, the chimney is a vertical columnar feature in a three-dimensional flow field that is only as wide perpendicular as it is parallel to natural flow.

The three-dimensional model builds on the Cannikin two-dimensional model by simply extending the domain in the direction of the island shoreline (perpendicular to the axes of the two-dimensional model). Thus, the finite-element mesh geometry of each vertical slice in the three-dimensional model is identical to the mesh geometry of the two-dimensional Cannikin model, with each element now having a constant width in the y direction (Figure 11).

The impacts of flow in a three-dimensional rubble chimney are simulated in a model 1,500 m wide, i.e., perpendicular to the natural flow direction (Figure 11). The chimney is simulated as a vertical column extending to ground surface that is rectangular in cross section and has a width of about two R_c, where R_c is the cavity radius (estimated to be 157 m). The hydraulic properties of the rock outside the chimney are considered to be

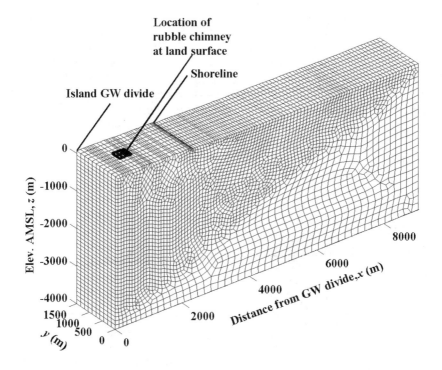

Figure 11: Mesh configuration used for simulations of the rubble chimney.

Parameter	Case #1	Case #2	Case #3
K_{xx}, K_{yy} (m/d)	1.86×10^{-2}	6.48×10^{-2}	1.78×10^{-2}
K_{zz} (m/d)	1.86×10^{-3}	6.48×10^{-3}	1.78×10^{-3}
K_{xx}, K_{yy}, K_{zz} of cavity and chimney (m/d)	1.86×10^{-2}	6.48×10^{-2}	1.78×10^{-2}
Rech (cm/yr)	6.13	3.33	1.89
θ_f	2.81×10^{-4}	2.71×10^{-4}	2.67×10^{-4}

Table 3: Values of parameters used in the three-dimensional rubble chimney simulations.

not significantly affected by the nuclear explosion and are assigned the background values of K and porosity. Model parameters that differ from the base-case Cannikin model for the three realizations are shown in Table 3.

The sensitivity studies are applied to three realizations selected out of the 260 runs for the Cannikin two-dimensional model. The parameter combinations of these realizations encompass a variety of positions of the TZ

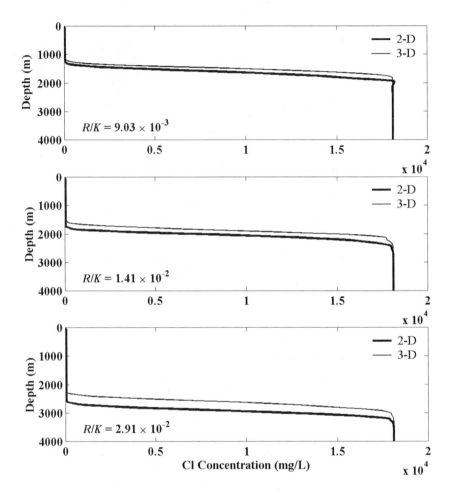

Figure 12: Comparison of vertical profiles of chloride concentration (mg/L) for two-dimensional and three-dimensional representations of the island hydraulics for the three selected realizations.

relative to the test cavity, while having virtually identical porosity (about 2.67×10^{-4}). Because the velocity field is very sensitive to porosity, this parameter was held constant to highlight the impact of the sensitivity cases. Though these realizations were selected from the realizations generated for Cannikin, the various positions of the TZ relative to the cavity allow them to represent flow fields possible for all three sites. The realization with the TZ well below the cavity is representative of Long Shot. The realizations with the cavity within and below the TZ are likely to be more representative of Milrow and Cannikin.

Simulation of the rubble chimney in three dimensions results in the simulation of a shallower TZ as compared to the two-dimensional case with the same parameter values (Figure 12). The magnitude of the difference is greatest for the highest R/K ratio, which places the TZ higher by about 500 m. Despite this, the test cavity remains well within the freshwater lens for this realization and thus flow velocities from the cavity are not impacted significantly. The inclusion of the three-dimensional rubble chimney for the lower R/K ratio places the TZ about 100 m higher, placing the cavity further into the low velocity saltwater zone.

The greater flux through the chimney and the radial flow in three dimensions introduce into the model a mechanism for lateral spreading of contaminants originating in the cavity that is not present in the two-dimensional model. The net effect is lower contaminant concentrations as the plume is diluted with a larger volume of groundwater. The result is that the two-dimensional model underestimates the effect of the chimney on slowing groundwater velocities and neglects dispersion in the third dimension. This result is true for all three R/K ratios and indicates that the use of the two-dimensional approximation for transport from the cavities is a conservative approach.

5.   CONCLUDING REMARKS

The uncertainty analysis for Amchitka Island not only provides information for a theoretical analysis of the importance and interplay among flow and transport parameters in coastal aquifers, it provides valuable information for managing this site of groundwater contamination. Uncertainty always exists when considering subsurface problems; quantifying the impact of that uncertainty on contaminant transport predictions can allow site managers to decide whether the uncertainty can be tolerated or must be reduced through additional data collection. Including the results of uncertainty (through a standard deviation on the breakthrough curves) always increases predicted transport. If a decision is made to reduce the uncertainty, the type of analysis shown here provides a quantitative framework for designing a field program with the highest chance of reducing model prediction uncertainty.

REFFERENCES

Armstrong, R.H., "Weather and climate," In: The Environment of Amchitka Island, Alaska, eds. M.L. Merritt and R.G. Fuller, 53–58, Energy Research and Development Administration, Technical Information Center, 1977.

Diersch, J.J., "Interactive, graphics-based finite-element simulation system FEFLOW for modeling groundwater flow contaminant mass and heat transport processes, FEFLOW Reference Manual," WASY Ltd., Berlin, 294 p., 1998.

Fenske, P.R., "Event-related hydrology and radionuclide transport at the Cannikin Site, Amchitka Island, Alaska," Desert Research Institute, Water Resources Center, Report 45001,NVO-1253-1, 41 p., 1972a.

Fenske, P.R., "Hydrology and radionuclide transport, Amchitka Island, Alaska," Desert Research Institute, Technical Report Series H-W, Hydrology and Water Resources Publication No. 12, 29 p., 1972b.

Green, G.W., "Some hydrological implications of temperature measurements in exploratory drillholes, Project Long Shot, Amchitka Island, Alaska," U.S. Geological Survey Technical Letter Goethermal—1, 8 p., 1965.

Hassan, A.E., Cushman, J.H. and Delleur, J. W., "Monte Carlo studies of flow and transport in fractal conductivity fields: Comparison with stochastic perturbation theory," *Water Resources Research*, **33**(11), 2519–2534, 1997.

Hassan, A.E., Cushman, J.H. and Delleur, J. W., "A Monte Carlo assessment of Eulerian flow and transport perturbation models," *Water Resources Research,* **34**(5), 1143–1163, 1998.

Hassan, A.E., Andricevic, R. and Cvetkovic, V., "Computational issues in the determination of solute discharge moments and implications for comparison to analytical solutions, *Advances in Water Resources*, **24,** 607–619, 2001.

LaBolle, E., Quastel, J., Fogg, G. and Gravner, J., "Diffusion processes in composite porous media and their integration by random walks: Generalized stochastic differential equations with discontinuous coefficients," *Water Resources Research,* **36**(3), 651–662, 2000.

LaBolle, E., Fogg, G. and Tompson, A.F.B., "Random-walk simulation of solute transport in heterogeneous porous media: Local mass-conservation problem and implementation methods," *Water Resources Research,* **32**(3), 583–593, 1996.

Pohlmann, K.F., Hassan, A.E. and Chapman, J.B., "Modeling density-driven flow and radionuclide transport at an underground nuclear test: Uncertainty analysis and effect of parameter correlation," *Water Resources Research*, **38**(5), 10.1029/2001WR001047, 2002.

Sass, J.H. and Moses, T.H., Jr., "Subsurface temperatures from Amchitka Island, Alaska," U.S. Geological Survey, Technical Letter, USGS 474-20 (Amchitka-16), 5 p., 1969.

Tompson, A.F.B. and Gelhar, L.W., "Numerical simulation of solute transport in three-dimensional, randomly heterogeneous porous media," *Water Resources Research,* **26**(10), 2451–2562, 1990.

Tompson, A.F.B., "Numerical simulation of chemical migration in physically and chemically heterogeneous porous media," *Water Resources Research,* **29**(11), 3709–3726,1993.

CHAPTER 11

Pumping Optimization in Saltwater-Intruded Aquifers

A.H.-D. Cheng, M.K. Benhachmi, D. Halhal, D. Ouazar, A. Naji,

K. EL Harrouni

1. INTRODUCTION

Coastal aquifers serve as major sources for freshwater supply in many countries around the world, especially in arid and semiarid zones. Many coastal areas are heavily urbanized, a fact that makes the need for freshwater even more acute [Bear and Cheng, 1999]. Inappropriate management of coastal aquifers may lead to the intrusion of saltwater into freshwater wells, destroying them as sources of freshwater supply. One of the goals of coastal aquifer management is to maximize freshwater extraction without causing the invasion of saltwater into the wells.

A number of management questions can be asked in such considerations. For existing wells, how should the pumping rate be apportioned and regulated so as to achieve the maximum total extraction? For new wells, where should they be located and how much can they pump? How can recharge wells and canals be used to protect pumping wells, and where should they be placed? If recycled water is used in the injection, how can we maximize the recovery percentage? These and other questions may be answered using the mathematical tool of optimization.

Efforts to improve the management of groundwater systems by computer simulation and optimization techniques began in the early 1970s [Young and Bredehoe, 1972; Aguado and Remson, 1974]. Since that time, a large number of groundwater management models have been successfully applied; see for example Gorelick [1983], Willis and Yeh [1987], and many other papers published in the *Journal of Water Resources Planning and Management, ASCE*, and the *Water Resources Research*. Applications of these models to aquifer situations with the explicit threat of saltwater intrusion in mind, however, are relatively few [Cumming, 1971; Cummings and McFarland, 1974; Shamir *et al.*, 1984; Willis and Finney, 1988; Finney *et al.*, 1992; Hallaji and Yazicigil, 1996; Emch and Yeh, 1998; Nishikawa,

1-56670-605-X/04/$0.00+$1.50
© 2004 by CRC Press LLC

1998; Das and Datta, 1999a, 1999b; Cheng *et al.*, 2000]. In terms of management objectives, some of these studies have addressed relatively complex settings such as mixed use of surface and subsurface water in terms of quantity and quality, water conveyance, distribution network, construction and utility costs, etc. However, saltwater intrusion into wells has been dealt with in simpler and indirect approaches, for example, by constraining drawdown or water quality at a number of control points, or by minimizing the overall intruded saltwater volume in the entire aquifer. The explicit modeling of saltwater encroachment into individual wells resulting in the removal of invaded wells from service is found only in Cheng *et al.* [2000].

This chapter reviews some of the earlier considerations of pumping optimization in saltwater-intruded aquifers under deterministic conditions, and furthermore, introduces the uncertainty factor into the management problem. The resultant methodology is applied to the case study of the City of Miami Beach in the northeast Spain.

2. DETERMINISTIC SIMULATION MODEL

The first step of modeling is to have a physical/mathematical model. Depending on the available data input from the field problem and the desirable outcome of the simulation, models of different levels of complexity, ranging from the sharp-interface model to the density-dependent miscible transport model, can be used [Bear, 1999]. For the method of solution, it can range from simple analytical solutions [Cheng and Ouazar, 1999] to the various finite-element- and finite-difference-based numerical solutions [Sorek and Pinder, 1999]. In principle, any of the above models and methods can be used; in reality, however, the selection of the model is dependent on the tolerable computer CPU time, as both the optimization and the stochastic modeling can be computational time consuming.

In our case, the Genetic Algorithm (GA) has been chosen as the optimization tool. Due to the large number of individual simulations needed in the GA, the simulation model needs to be highly efficient in order to stay within a reasonable amount of computation time. For this reason, the sharp interface analytical solution is chosen, which is briefly described in the following.

Figures 1(a) and (b) respectively give the definition sketch of a confined and an unconfined aquifer. The aquifers are with homogeneous hydraulic conductivity K and constant thickness B in the confined aquifer case. Distinction has been made between two zones—a freshwater only zone (zone 1), and a freshwater–saltwater coexisting zone (zone 2). Following the

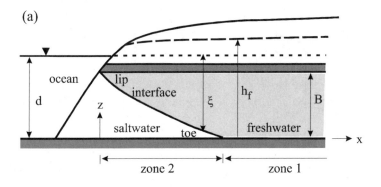

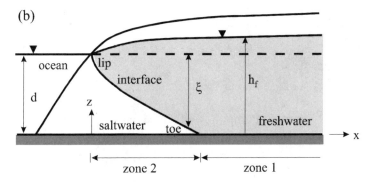

Figure 1: Definition sketch of saltwater intrusion in (a) a confined aquifer, and (b) an unconfined aquifer.

work of Strack [1976], the Dupuit-Forchheimer hydraulic assumption is used to vertically integrate the flow equation, reducing the solution geometry from three-dimensional to two-dimensional (horizontal x-y plane). Steady state is assumed. The Ghyben-Herzberg assumption of stagnant saltwater is utilized to find the saltwater–freshwater interface. With the above common assumptions of groundwater flow, the governing equation for the system is the Laplace equation:

$$\nabla^2 \phi = 0 \tag{1}$$

where ∇^2 is the Laplacian operator in two-spatial dimensions (x and y), and the potential ϕ is defined differently in the two zones

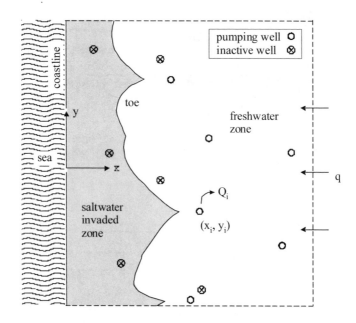

Figure 2: Pumping wells in a coastal aquifer.

$$\phi = Bh_f = \frac{1}{2(s-1)}[h_f + (s-1)B - sd]^2 \quad \text{for zone 1}$$

$$\phi = \frac{1}{2(s-1)}[h_f + (s-1)B - sd]^2 \quad \text{for zone 2}$$

(2)

for confined aquifer; and

$$\phi = \frac{1}{2}[h_f^2 - sd^2] \quad \text{for zone 1}$$

$$\phi = \frac{s}{2(s-1)}(h_f - d)^2 \quad \text{for zone 2}$$

(3)

for unconfined aquifer. We also define

$$s = \frac{\rho_s}{\rho_f}$$

(4)

as the saltwater and freshwater density ratio, and other definitions are found in Figure 1.

In our problem, we consider a semi-infinite coastal plain bounded by a straight coastline aligned with the y-axis (Figure 2). Multiple pumping wells are located in the aquifer with coordinates (x_i, y_i) and discharge Q_i. There is a uniform freshwater outflow rate q. The aquifer can be confined or unconfined. Solution of the potential ϕ for this problem can be found by the method of images and has been given by Strack [1976] (see also Cheng and Ouazar, 1999):

$$\phi = \frac{q}{K}x + \sum_{i=1}^{n}\frac{Q_i}{4\pi K}\ln\left[\frac{(x-x_i)^2 + (y-y_i)^2}{(x+x_i)^2 + (y-y_i)^2}\right] \tag{5}$$

With the above solution, the toe location of saltwater wedge x^{toe} is found where the potential takes the value ϕ^{toe},

$$\phi^{toe} = \frac{q}{K}x^{toe} + \sum_{i=1}^{n}\frac{Q_i}{4\pi K}\ln\left[\frac{(x^{toe}-x_i)^2 + (y-y_i)^2}{(x^{toe}+x_i)^2 + (y-y_i)^2}\right] \tag{6}$$

where

$$\phi^{toe} = \frac{s-1}{2}B^2 \quad \text{for confined aquifer}$$

$$\phi^{toe} = \frac{s(s-1)}{2}d^2 \quad \text{for unconfined aquifer} \tag{7}$$

Since ϕ^{toe} is some known number evaluated from Eq. (7), Eq. (6) can be solved for x^{toe} for each given y value using a root finding technique.

3. OPTIMIZATION UNDER DETERMINISTIC CONDITIONS

The management objective of the coastal pumping operation is to maximize the economic benefit from the pumped water less the utility cost for lifting the water. For simplicity, we assume that the value of water and the utility cost are both linear functions of discharge Q_i. The objective is to maximize the benefit function Z with respect to the design variables Q_i [Haimes, 1977]:

$$\max_{Q_i} Z = \sum_{i=1}^{n}Q_i\left[B_p - C_P\left(L_i - h_i\right)\right] \tag{8}$$

In the above B_p is the economic benefit per unit discharge, C_p is the cost per unit discharge per unit lift height, L_i is the ground elevation at well i, and h_i is the water level in well i. It should be remarked that although a relative simple model is used for the right-hand side of Eq. (8), it can be generalized to a realistic microeconomic model involving supply and demand without complicating the solution process.

The pumping operation is subject to some constraints. First, the discharge of each well must stay within the certain limits set by the operation conditions such as the minimum feasible pumping rate, maximum capacity of the pump, restriction on well drawdown, etc. This can be written as

$$Q_i^{\min} \leq Q_i \leq Q_i^{\max} \quad \text{or} \quad Q_i = 0; \quad \text{for} \quad i = 1, \ldots, n \tag{9}$$

We note that the second condition in the above allows the well to be shut down. Second, it is required that saltwater wedge does not invade the pumping wells

$$x_i^{toe} < x_i \quad \text{at} \quad y = y_i; \quad \text{for all active wells} \tag{10}$$

where x_i^{toe} stands for the toe location in front of well i.

Since genetic algorithm can only work with unconstrained problems, it is necessary to convert the constrained problem described by Eqs. (8)-(9) to an unconstrained one. This is accomplished by the adding penalty to the objective function for any violation that takes place:

$$\max_{Q_i} \ Z = \sum_{i=1}^{n} Q_i \left[B_p - C_P \left(L_i - h_i \right) \right] - r_i N_i \left(\frac{x_i^{toe}}{x_i} - 1 \right)^2 \tag{11}$$

where r_i are penalty factors, which are empirically selected, and $N_i = 1$ for $x_i^{toe} \geq x_i$ and $N_i = 0$ for $x_i^{toe} < x_i$. We notice that the constraint Eq. (9) is not included in Eq. (11) because it is automatically satisfied by setting the population space in genetic algorithm.

4. GENETIC ALGORITHM

Conventional optimization techniques, such as the linear and nonlinear programming, and gradient-based search techniques are not suitable for finding global optimum in space that is discontinuous and contains a large number of local optima, which are the prevalent conditions for the optimization problem defined above. To overcome these difficulties, a genetic algorithm (GA) has been introduced and successfully applied [Cheng et al., 2000]. GA is a probabilistic search based optimization technique that imitates the biological process of evolution [Holland, 1975]. Its application to groundwater problems started in the mid-1990s [McKinney and Lin, 1994; Ritzel et al., 1994; Rogers and Fowla, 1994; Cienlawski et al., 1995], and since that time it has found many applications. (See Ouazar and Cheng [1999] for a review.)

A brief illustration of the GA solution procedure applied to the current problem is given below. Given the solution space of Q_i defined by Eq. (9), we discretize it in order to reduce the number of trial solutions from infinite to a finite set. As an example, if each discharge is constrained between $100 \leq Q_i \leq 500$ m³/day, and the desirable accuracy of the solution is 5 m³/day (which is a rather crude resolution), then for each Q_i there exist 82 possible discrete values (including the zero pumping rate). If there are 10 wells in the field, then the total number of possible combinations of pumping rate is $82^{10} = 1.4 \times 10^{19}$. One of the combinations is the optimal pumping solution we look for. This search space is so huge that if we spend 1 sec of CPU time to conduct a single simulation to check its benefit, it will take 4×10^{11} years to complete the work. The search space of a typical field problem in fact is greater than the above. Hence we must follow some intelligent rules in the search; this is where the GA comes in.

GA seeks to represent the search space by binary strings. In the above example, it is sufficient to represent all possible combinations of pumping rate by a 64-bit binary string ($2^{64} = 1.8 \times 10^{19}$). To seed an initial population, a random number generator is used to flip the bits between 0 and 1 to create individuals in the form of 01101...10111 (64 digits long), each one corresponding to a distinct set of pumping rates. Typically a relatively small number of individuals, say 10 to 20, are created to fill a generation. Individuals are then tested for their fitness to survive by running the deterministic simulation as described above. The fitness is determined by the objective function given as the right-hand side of Eq. (11).

Once the fitness is determined for each individual in the generation, certain evolutional-based probabilistic rules are applied to breed better offspring. For example, in a simple genetic algorithm (SGA), three rules, *selection*, *crossover*, and *mutation*, are used [Michalewicz, 1992]. First, the *selection* process decides whether an individual will survive by "throwing a dice" using a probability proportional to the individual's fitness value. Second, the GA disturbs the resulting population by performing *crossover* with a probability of p_c. In this operation, each binary string (individual) is considered as a chromosome. Segments of chromosome between individuals can be exchanged according to the predetermined probability. Third, to create diversity of the solution, GA further perturbs the population by performing *mutation* with a probability of p_m. In this operation, each bit of the chromosome is subjected to a small probability of mutation by allowing it to be flipped from 1 to 0 or the other way around. After these steps, a new generation is formed and the evolution continues. The process is terminated

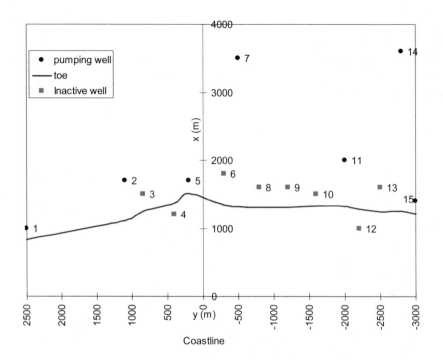

Figure 3: Pumping wells in a coast and saltwater intrusion front.

by a number of criteria, such as no improvement observed in an number of generations, or reaching a pre-determined maximum number of generation. The reader can consult the above-cited references for more detail.

5. EXAMPLE OF DETERMINISTIC OPTIMIZATION

This test case was examined in Cheng *et al.* [2000]. Assume an unconfined aquifer with $K = 40$ m/day, q = 40 m²/day, d = 15 m, ρ_s = 1.025 g/cm³, and ρ_f = 1 g/cm³. Figure 3 gives an aerial view of the coast and the locations of 15 pumping wells. The well coordinates are shown in columns (2) and (3) of Table 1. Each well is bounded by a maximum and a minimum well discharge, as indicated in columns (4) and (5). In this optimization problem, only the benefit from the pumped volume is considered, and the utility cost is neglected. The objective function (11) is modified to

$$\max_{Q_i} Z = \sum_{i=1}^{n} Q_i - r_i N_i \left(\frac{x_i^{toe}}{x_i} - 1 \right)^2 \qquad (12)$$

(1)	(2)	(3)	(4)	(5)	(6)	(7)
Well Id	x_i (m)	y_i (m)	Q_i^{max} (m³/day)	Q_i^{min} (m³/day)	Q_i (m³/day)	x_i^{toe} (m)
1	1000	2500	600	150	201	836
2	1700	1100	1300	150	351	1117
3	1500	850	1100	150	0	1257
4	1200	400	800	150	0	1372
5	1700	200	1300	150	150	1514
6	1800	-300	1400	150	0	1344
7	3500	-500	1500	150	1497	1323
8	1600	-800	1200	150	0	1311
9	1600	-1200	1200	150	0	1315
10	1500	-1600	1100	150	0	1332
11	2000	-2000	1500	150	155	1319
12	1000	-2200	600	150	0	1287
13	1600	-2500	1200	150	0	1241
14	3600	-2800	1500	150	1387	1251
15	1400	-3000	1000	150	150	1213
Total					3891	

Table 1: Optimal pumping well solution.

The GA described earlier is used for optimization. In the first attempt, the optimization was conducted by assuming all 15 wells are in operation. The search space for each well is defined between Q_i^{min} and Q_i^{max} with increment size of roughly 1 m³/day and also the zero discharge. If a well is invaded, a penalty is imposed with an empirical penalty factor r_i to discourage such events. If the well is shut down, $Q = 0$, the program detects it and no penalty is applied for invasion. This allows the inactive wells to be intruded in order to increase pumping.

After three runs of GA with different seeding of initial population, the best solution gives the total discharge of 3,610 m³/day. The optimal solution shows that eight wells are in operation and seven are shut down. The fact that so many wells are shut down is not surprising, as an estimate based on a simple analytical solution [Cheng et al., 2000] shows that the well field is too crowded and some wells can be taken out of action.

The program was run on a Pentium 450MHz microcomputer. It was terminated when the maximum number of generations was reached, for about 6 hours of CPU time. Since an near optimal solution may not have been reached, a second search is conducted using a refined strategy. In the second search, only cases with any combinations of seven, eight, and nine wells in operation are admitted into the search space. Wells not selected do not exist and can be invaded. This strategy much reduces the size of the search space and better solution is obtained. The best solution is a seven-well case as shown in column (6) of Table 1. The toe location in front of the wells is shown in column (7). The total pumping rate is 3,891 m³/day. The saltwater intrusion front is graphically demonstrated in Figure 3, with the well locations marked. We notice that two of the inactive wells, 4 and 12, are intruded by saltwater.

6. STOCHASTIC SIMULATION MODEL

The solution presented above assumes deterministic conditions, i.e., all aquifer data are known with certainty. This is not true in reality as hydrogeological surveys are expensive and time consuming to conduct; hence hydrogeological data are rare. The optimization model needs to take this reality into consideration.

The first step of conducting a stochastic optimization is to have a stochastic simulation model. This can be accomplished by applying the second order uncertainty analysis of Cheng and Ouazar [1995] to the deterministic model given as Eq. (6). Based on the approximation of Taylor series, the statistical moments of toe location can be related to the moments of uncertain parameters as [Naji et al., 1998]

$$\overline{x}^{toe} = x^{toe}\left(\overline{q},\overline{K}\right) + \frac{1}{2}\left[\frac{\partial^2 x^{toe}}{\partial q^2}\sigma_q^2 + \frac{\partial^2 x^{toe}}{\partial K^2}\sigma_K^2\right] \tag{13}$$

$$\sigma_x^2 = \left(\frac{\partial x^{toe}}{\partial q}\right)^2 \sigma_q^2 + \left(\frac{\partial x^{toe}}{\partial K}\right)^2 \sigma_K^2 \tag{14}$$

where $\overline{x}^{toe}$, $\overline{q}$, and $\overline{K}$ are respectively the mean toe location, the mean freshwater outflow rate, and the mean hydraulic conductivity; σ_x^2, σ_q^2, and σ_K^2 are respectively the variance of toe location, freshwater outflow rate, and hydraulic conductivity; and $x^{toe}\left(\overline{q},\overline{K}\right)$ is the toe location evaluated using the mean parameter values. In the above, we have neglected the covariance σ_{qK} by assuming that it is small. The above equations state that in order to obtain the mean toe location and its standard deviation, we first need to calculate the toe location using the mean parameter values, i.e., $x^{toe}\left(\overline{q},\overline{K}\right)$. This is obtained from the deterministic solution by solving Eq. (6) using the given $\overline{q}$ and $\overline{K}$ values. Next, we need to find the partial derivatives of toe location with respect to q and K. This is found by perturbing the q and K values by small amounts in Eq. (6). In other words, Eq. (6) is solved for the toe location using values of $\overline{q} \pm \Delta q$ and $\overline{K} \pm \Delta K$ and the difference in x^{toe} is found. Utilizing finite difference approximation, the partial derivative $\partial x^{toe} / \partial q$, $\partial^2 x^{toe} / \partial q^2$, etc., can be approximated. Given the variances of aquifer data, σ_q^2 and σ_K^2, we can then assemble the mean toe location and its standard deviation from Eqs. (13) and (14). More detail of the above procedure can be found in Cheng and Ouazar [1995], and Naji et al. [1998, 1999].

7. CHANCE CONSTRAINED OPTIMIZATION

The optimization problem described in Sections 2 through 5 is based on deterministic conditions. In the event of input data uncertainty, a stochastic optimization is necessary. The chance-constrained programming [Charnes and Cooper, 1959; 1963] is used for this purpose. This optimization model allows us to use stochastic parameters as input data and produces an output prediction based on desirable reliability level.

Charnes and Cooper [1959, 1963] studied chance constrained programming by transforming a stochastic optimization problem into a deterministic equivalent. The chance-constrained programming can incorporate reliability measures imposed on the decision variables. This methodology has been applied to solve a number of groundwater management problems. Tung [1986] developed a chance-constrained model that takes into account the random nature of transmissivity and storage coefficient. Wagner and Gorelick [1987] presented a modified form of the chance constrained programming to determine a pumping strategy for controlling groundwater quality. Hantush and Marino [1989] presented a chance-constrained model for stream-aquifer interaction. Morgan et al. [1993] developed a mixed-integer chance-constrained programming and demonstrated its applicability to groundwater remediation problems. Chance-constrained groundwater management models have also been applied to design groundwater hydraulics [Tiedman and Gorelick, 1993] and quality management strategies [Gailey and Gorelick, 1993]. Chan [1994] developed a partial infeasibility method for aquifer management. Datta and Dhiman [1996] utilized a chance-constrained model for designing a groundwater quality monitoring network. Wagner [1999] employed the chance-constrained model for identifying the least cost pumping strategy for remediating groundwater contamination. Sawyer and Lin [1998] considered the combination of uncertainty in the cost coefficients and constraints of the groundwater management model.

For the present problem we assume that the freshwater outflow rate q and the hydraulic conductivity K are random variables, causing the toe location in front of each well x_i^{toe} to be uncertain. The constraint given by Eq. (10) needs to be modified to a probabilistic one:

$$\text{Prob}\left(x_i^{toe} < x_i\right) \geq R; \quad \text{for all active wells} \tag{15}$$

where R is the desirable reliability level of prediction set by the water manager. The chance constraint converts the above probabilistic constraint into a deterministic one:

$$\overline{x}_i^{toe} + F^{-1}(R)\sigma_{x_i^{toe}} < x_i; \quad \text{for all active wells} \tag{16}$$

where $\overline{x}_i^{toe}$ is the expectation and $\sigma_{x_i^{toe}}$ is the standard deviation of the toe location x_i^{toe}, and $F^{-1}(R)$ is the value of the standard normal cumulative probability distribution corresponding to the reliability level R. The chance-constrained optimization problem is then defined by the objective function Eq. (8), which is subject to the constraints Eqs. (9) and (16).

In order to apply GA for the solution of the optimization problem, we need to convert the constrained problem to an unconstrained one. Similar to the deterministic problem, this is accomplished by imposing penalty for the violation of the chance constraint Eq. (16):

$$\max_{Q_i} Z = \sum_{i=1}^{n} Q_i \left[B_p - C_P \left(L_i - h_i \right) \right] - r_i N_i \left(\frac{\overline{x}_i^{toe} + F^{-1}(R)\sigma_{x_i^{toe}}}{x_i} - 1 \right)^2 \tag{17}$$

which can be compared to its deterministic counterpart Eq. (11). The GA methodology as described in Section 4 is then applied for its solution.

8.   CASE STUDY—MIAMI BEACH, SPAIN

The above-proposed optimization model has been tested and applied to a few hypothetical as well as real cases [Benhachmi *et al.*, 2003a, b]. Here, we report the case study of the city of Miami Beach in northeast Spain.

A large fraction of the total population of Spain (about 80% of its 6 million inhabitants) lives along the Catalonia coast [Bayó *et al.*, 1992]. This concentration of population creates large freshwater demands for domestic consumption, in addition to the agricultural, industrial, and tourism needs. Aquifers along the coast have been subjected to intensive exploitation;

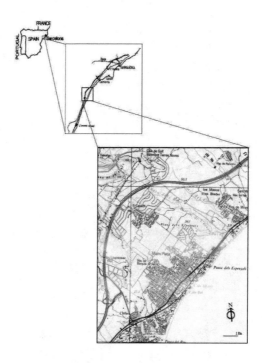

Figure 4: Location of Miami Beach, Spain.

consequently, excessive salinity in well water is a common occurrence [Bayó *et al.*, 1992; Himmi, 2000]. In many situations, there is a poor understanding of aquifer response, detailed studies are lacking, and the monitoring of seawater intrusion is insufficient. In spite of the strict regulations introduced in the Water Act of Spain, control of abstractions is scarce. In the coastal area of Tarragona, north to Ebre, saltwater intrusion is caused by the concentrated abstraction near the coast, which has contaminated many wells and forced the freshwater importation of up to 4 m³/s from the Ebre river by means of an 80 km canal and pipeline.

The current situation is in part a result of inadequate water resources planning and management. The unfortunate consequence of management failure is that there generally exists distrust in the public in the feasibility of using coastal groundwater resources to meet water demands, and solutions that need large amounts of investment are rejected. However, it is believed that with adequate management and enforcement, some of the current problems can be alleviated.

In the present work, we shall apply the previously described stochastic optimization approach to the management of the Miami

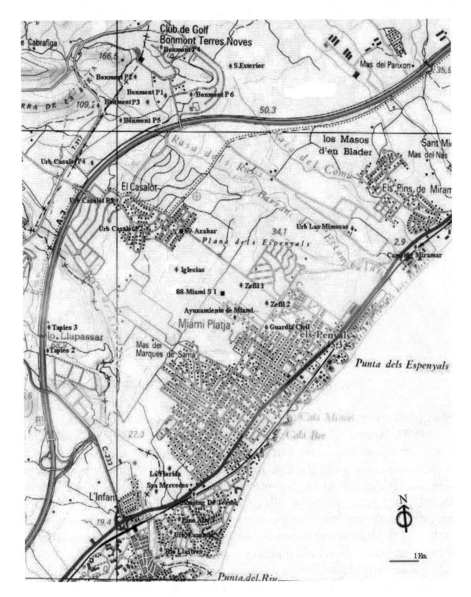

Figure 5: City of Miami Beach, Spain, and pumping well locations.

unconfined aquifer located near Tarragona, Spain (Figure 4). For many years, the aquifer has been one of the most important water-supply sources for the city of Miami Beach for domestic purposes. The study area is located southwest of the city of Tarragona and encompasses about 17 km². Lithology of Miami aquifer consists of unconsolidated sediments of Quaternary

(1) No.	(2) Well Id	(3) x_w (m)	(4) y_w (m)	(5) Q_{max} (m³/day)	(6) Q_{min} (m³/day)	(7) L_i (m)
1	Bonmont P4	3877	4362	1200	120	80
2	Bonmont P2	3826	3748	1200	120	113
3	Bonmont P5	3655	3390	1200	120	111
4	Urb. Casalot P4	3625	2648	1200	120	89
5	Bonmont P3	3507	3686	1200	120	81
6	Bonmont P1	3469	3900	1200	120	78
7	Bonmont P6	3285	4148	1200	120	66
8	S. Exterior	3161	4715	1200	120	67
9	Urb. Casalot P3	3133	2593	1200	120	85
10	Tapies 3	2808	961	1200	120	91
11	Urb. Casalot P2	2744	2705	1200	120	70
12	Tapies 2	2647	759	1200	120	89
13	Iglesias	2047	2496	1200	120	65
14	Zefil 1	1322	2922	1200	120	25
15	Ayu. De Miami	1246	2541	1200	120	30
16	Zefil 2	1077	2769	1200	120	22
17	Guardia Civil	906	2761	1200	120	19
18	Urb. Las Mimosas	873	4202	1200	120	20
19	La Florida	704	763	1200	120	34
20	Pozo de Sra. Mercedes	431	677	1200	120	20
21	C. Terme	358	672	1200	120	15
22	C. Miramar	304	4564	1200	120	12
23	Pino Alto 3	244	399	1200	120	13
24	Urb. Euromar	206	315	1200	120	14
25	Rio Llastres	179	101	1200	120	12

Table 2: Pumping well locations and discharge limits for the Miami Beach aquifer.

age, corresponding to coastal piedmonts and alluvial fans, and is generally unconfined and single-layered. The sediment consists of clay and gravel, and overlies a blue clay of Pliocene age, which constitutes the effective lower hydrologic boundary.

The unconfined aquifer of Miami Beach is examined. Its hydraulic parameters are estimated to be: mean hydraulic conductivity $\overline{K}$ = 14 m/day, mean freshwater outflow rate $\overline{q}$ = 1.2 m³/day/m, average aquifer thickness d

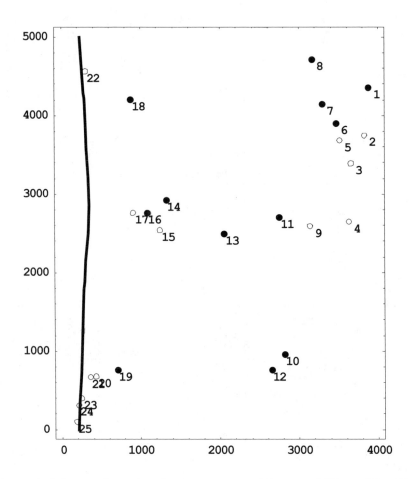

Figure 6: Saltwater intrusion for the case $c_q = 5\%$, $c_K = 25\%$ and $R = 90\%$.
(Solid circle: active well; open circle: inactive well.)

$= 30$ m, and densities of freshwater and saltwater are $\rho_f = 1.0$ g/cm³ and
$\rho_s = 1.025$ g/cm³. To calculate the benefit as defined in Eq. (17), we use
0.01€ per m³ for the uniform benefit rate for water produced, and 0.0002 €
per m³ of water per m pumping lift for the utility cost. Taking into
consideration that the information about freshwater outflow rate and
hydraulic conductivity is uncertain, we further estimate that the coefficients
of variation for these quantities are $c_q = \sigma_q / \overline{q} = 5\%$ and $c_K = \sigma_K / \overline{K} = 25\%$.
In the chance-constrained model, the final result is dependent on the required

Well	Well Discharge (m³/day)				
	Case 1 C_K=25% C_q=5% R=90%	Case 2 C_K=25% C_q=5% R=95%	Case 3 C_K=25% C_q=5% R=99%	Case 4 C_K=1% C_q=5% R=95%	Case 5 C_K=50% C_q=5% R=95%
1	724	539	962	0	474
2	0	0	0	231	0
3	0	219	0	0	0
4	0	0	396	584	338
5	0	0	0	323	988
6	700	755	209	580	408
7	1189	978	929	1011	350
8	797	932	685	771	809
9	0	751	0	628	0
10	386	191	310	0	291
11	661	0	296	899	230
12	651	142	307	127	277
13	913	539	506	408	549
14	265	214	257	268	214
15	0	184	0	281	205
16	238	229	263	210	177
17	0	285	283	220	0
18	179	139	283	255	141
19	215	287	291	145	171
20	0	0	0	0	0
21	0	0	0	0	0
22	0	0	0	0	0
23	0	0	0	0	0
24	0	0	0	0	0
25	0	0	0	0	0
Total	6918	6384	5977	6941	5622

Table 3: Optimal pumping pattern for various input data uncertainty and output prediction reliability levels.

reliability—the higher the reliability required, the lower the extraction rate. Here we choose $R = 90\%$. These complete the data input requirements for the stochastic optimization problem.

Figure 5 gives an aerial view of the coast and the locations of 25 pumping wells in the aquifer. The well coordinates are shown in columns 3

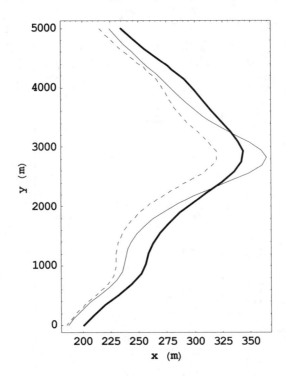

Figure 7: Saltwater intrusion front (exaggerated scale in x-direction). (Thick solid line: case 1, $R = 90\%$; thin solid line: case 2, $R = 95\%$; dash line: case 3, $R = 99\%$.)

and 4 in Table 2, which are ranked by their distance to the coast. For each well, a lower bound pumping rate $Q_i^{\min}$ and an upper bound $Q_i^{\max}$ are given, as shown in columns 5 and 6. Column 7 shows the ground elevation of the well.

The GA is utilized for the search of a near optimal solution. The following parameters are used in the GA simulation: population size = 20, maximum number of generations = 200. Different values of crossover and mutation probabilities are used during the testing phase. For results presented here, $p_c = 0.7$ and $p_m = 0.1$ are used.

Since the search space is large, some manual intervention is used to assist in the optimization. First, by visual inspection, it is clear that the six wells numbered 20 to 25 (Figure 6) are too close to the coast. These wells are manually shut down, meaning that they are not in the search space and saltwater is readily allowed to invade. This action will permit the inland

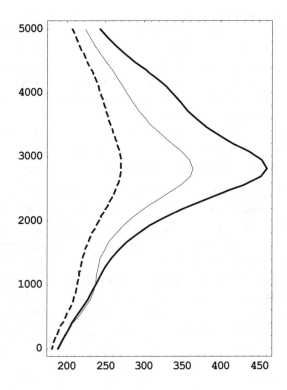

Figure 8: Saltwater intrusion front (exaggerated scale in x-direction). (Thick solid line: case 4, $c_K = 1\%$; thin solid line: case 2, $c_K = 25\%$; dash line: case 5, $c_K = 50\%$.)

wells to pump more. The next decision comes to the well group 14 to 17 (see Figure 6), whether they can be shut down as well. These are an important municipal group supplying for domestic consumption; hence heavy penalty is imposed for their invasion.

The resultant pumping pattern for the current case of $c_q = 5\%$, $c_K = 25\%$ and $R = 90\%$ is shown in Table 3 as case 1. We observe that in addition to wells 20 to 25, which are manually shut down, some other wells are shut down as well as the result of GA simulation. The total pumping rate is 6,918 m³/day. The resultant mean saltwater intrusion front is shown in Figure 6. Figure 6 also marks the well locations and numbers, with open circles indicating wells that are shut down, and solid circles for wells in operation.

In the next simulation, case 2, we fix the input data uncertainty, but change the required output reliability to a higher number $R = 95\%$. The resultant pumping pattern is shown as case 2 in Table 3. We observe that the total pumping rate is decreased to 6,384 m^3/day. If we further increase the reliability to $R = 99\%$, the optimal pumping rate is further reduced to 5,977 m^3/day, as shown in case 3 of Table 3. To show the difference in the mean saltwater intrusion front, the three cases are plotted in Figure 7. We observe that the mean saltwater intrusion front is more receded toward the coast to allow for high reliability of prediction.

Next, we examine the effect of data uncertainty. In cases 4 and 5, we fix the reliability level to $R = 95\%$, same as case 2. For case 4, we use the same coefficient of variation for freshwater outflow rate, $c_q = 5\%$, but assume that the hydraulic conductivity is known with high precision, $c_K = 1\%$. The simulated result is shown in Table 3, which gives the total well discharge as 6,941 m^3/day, larger than the value of 6,384 m^3/day for case 2. Hence reducing the data uncertainty of the input data can increase the allowable pumping rate. In the next case, we keep all data the same except that c_K is changed to 50%. The resultant pumping rate is shown as case 5 in Table 3, with the total pumping rate 5,622 m^3/day. So the increased data uncertainty has caused a reduction in allowable pumping. The mean saltwater intrusion front of the three cases, 2, 4, and 5 are shown in Figure 8 for comparison.

9. CONCLUSION

In this chapter we presented an optimization model for maximizing the benefit of pumping freshwater from a group of coastal wells under the threat of saltwater invasion. In view of the real-world situation, the aquifer properties are assumed to be uncertain, and are given in terms of mean values and standard deviations. The predicted maximum pumping rate is dependent on the desirable reliability that can be specified by the manager. The tools used in the optimization problem include analytical solution of sharp interface model, the stochastic solution based on perturbation, the chance-constrained programming, and the genetic algorithm.

The simulations based on the data of Miami Beach, Spain, show that the reduced aquifer data uncertainty can increase the economic benefit by pumping more water. To reduce input data uncertainty, however, hydrogeological studies need to be conducted, which involve certain costs. The trade-offs between increased benefit from pumping and the cost of data gathering can also be modeled into the objective function. This is however not attempted in this chapter.

The results show that the desirable reliability of prediction can also affect the allowable pumping rate. The higher the reliability, the lower the amount of water that can be pumped. The choice of reliability is dependent on the costs of the failure of the system—what will be the cost of loss of water, the cost of restoration, and any environmental consequences? These factors can also be programmed into the objected function if these costs can be estimated.

In conclusion, we shall emphasize that a strict deterministic prediction is non-conservative and is prone to failure. To guard against failure, a safety factor, which is typically arbitrary, can be imposed. A too conservative safety factor causes waste, and a non-conservative one may not be safe. The stochastic optimization procedure presented in this chapter offers a rational and optimal way to approach the uncertainty problem. The coastal water managers can weigh factors such as investing money to gather aquifer data to raise confidence level, pumping more and risking failure if an alternative source of water is available, the long-term and short-term economical projections, the environmental consequences, etc., to make the best decision based on the information available.

REFERENCES

Aguado, E. and Remson, I., "Ground-water hydraulics in aquifer management," *J. Hyd. Div., ASCE,* **100**, 103–118, 1974.

Bayó, A, Loaso, C., Aragones, J.M. and Custodio, E., "Marine intrusion and brackish water in coastal aquifers of southern Catalonia and Castello (Spain): A brief survey of actual problems and circumstances," Proc. 12th Saltwater Intrusion Meeting, Barcelona, 741–766, 1992.

Bear, J., "Conceptual and mathematical modeling," Chap. 5, In: *Seawater Intrusion in Coastal Aquifers—Concepts, Methods, and Practices,* eds. J. Bear, A.H.-D. Cheng, S. Sorek, D. Ouazar and I. Herrera, Kluwer, 127–161, 1999.

Bear, J. and Cheng, A.H.-D., "An overview," Chap. 1, In: *Seawater Intrusion in Coastal Aquifer—Concepts, Methods, and Practices*, eds. J. Bear, A.H.-D. Cheng, S. Sorek, D. Ouazar and I. Herrera, Kluwer, 1–8, 1999.

Benhachmi, M.K., Ouazar, D., Naji, A., Cheng, A.H.-D. and EL Harrouni, K., "Pumping optimization in saltwater intruded aquifers by simple genetic algorithm—Deterministic model," Proc. 2nd Int. Conf. Saltwater Intrusion and Coastal Aquifers—Monitoring, Modeling, and Management, Merida, Mexico, March 30–April 2, 2003a.

Benhachmi, M.K., Ouazar, D., Naji, A., Cheng, A.H.-D. and EL Harrouni, K., "Pumping optimization in saltwater intruded aquifers by simple genetic algorithm—Stochastic model," Proc. 2nd Int. Conf. Saltwater Intrusion and Coastal Aquifers—Monitoring, Modeling, and Management, Merida, Mexico, March 30–April 2, 2003b.

Chan, N., "Partial infeasibility method for chance-constrained aquifer management," *J. Water Resour. Planning Management, ASCE*, **120**, 70–89, 1994.

Charnes, A. and Cooper, W.W., "Chance-constrained programming," *Mgmt. Sci.*, **6**, 73–79, 1959.

Charnes, A. and Cooper, W.W., "Deterministic equivalents for optimizing and satisfying under chance constraints," *Oper. Res.*, **11**, 18–39, 1963.

Cheng, A.H.-D., Halhal, D., Naji, A. and Ouazar, D., "Pumping optimization in saltwater-intruded coastal aquifers," *Water Resour. Res.*, **36**, 2155–2166, 2000.

Cheng, A.H.-D. and Ouazar, D., "Theis solution under aquifer parameter uncertainty," *Ground Water*, **33**, 11–15, 1995.

Cheng, A.H.-D. and Ouazar, D., "Analytical solutions," Chap. 6, In: *Seawater Intrusion in Coastal Aquifers—Concepts, Methods, and Practices*, eds. J. Bear, A.H.-D. Cheng, S. Sorek, D. Ouazar and I. Herrera, Kluwer, 163–191, 1999.

Cienlawski, S.E., Eheart, J.W. and Ranjithan, S., "Using genetic algorithms to solve a multiobjective groundwater monitoring problem," *Water Resour. Res.*, **31**, 399–409, 1995.

Cummings, R.G., "Optimum exploitation of groundwater reserves with saltwater intrusion", *Water Resour. Res.*, **7**, 1415–1424, 1971.

Cummings, R.G. and McFarland, J.W., "Groundwater management and salinity control," *Water Resour. Res.*, **10**, 909–915, 1974.

Das, A. and Datta, B., "Development of multiobjective management models for coastal aquifers," *J. Water Resour. Planning Management*, *ASCE*, **125**, 76–87, 1999a.

Das, A. and Datta, B., "Development of management models for sustainable use of coastal aquifers," *J. Irrigation Drainage Eng.*, *ASCE*, **125**, 112–121, 1999b.

Datta, B.D. and Dhiman, S.D., "Chance-constrained optimal monitoring network design for pollutants in ground water," *J. Water Resour. Planning Management, ASCE*, **122**, 180–188, 1996.

Emch, P.G. and Yeh, W.W.G., "Management model for conjunctive use of coastal surface water and groundwater," *J. Water Resour. Planning Management, ASCE*, **124**, 129–139, 1998.

Finney, B.A., Samsuhadi and Willis, R., "Quasi-3-dimensional optimization model of Jakarta Basin," *J. Water Resour. Planning Management*, *ASCE*, **118**, 18–31, 1992.

Gailey, R.M. and Gorelick, S.M., "Design of optimal, reliable plume capture schemes: Application to the Gloucester landfill groundwater contamination problem," *Ground Water*, **31**, 107–114, 1993.

Gorelick, S.M., "A review of distributed parameter groundwater management modeling methods," *Water Resour. Res.*, **19**, 305–319, 1983.

Haimes, Y.Y., *Hierarchical Analyses of Water Resources Systems*, McGraw-Hill, 1977.

Hallaji, K. and Yazicigil, H., "Optimal management of coastal aquifer in Southern Turkey," *J. Water Resour. Planning Management, ASCE*, 122, 233–244, 1996.

Himmi, M., "Délimitacion de la intrusion marina en los acuiferos costeros por metodos geofisicos," doctoral dissertation, Universidad de Barcelona, Facultad de Géologia, 2000.

Holland, J., *Adaptation in Natural and Artificial Systems*, Univ. Michigan Press, Ann Arbor, 1975.

Hantush, M.M.S. and Marino, M.A., "Chance-constrained model for management of stream-aquifer system," *J. Water Resour. Planning Management, ASCE*, **115**, 259–277, 1989

McKinney, D.C. and Lin, M.D., "Genetic algorithm solution of groundwater-management models," *Water Resour. Res.*, **30**, 1897–1906, 1994.

Michalewicz, Z., *Genetic Algorithms + Data Structures = Evolution Programs*, Springer-Verlag, 1992.

Morgan, D.R., Eheart, J.W. and Valocchi, A.J., "Aquifer remediation design under uncertainty using a new chance constrained programming technique," *Water Resour. Res.*, **29**, 551–561, 1993.

Naji, A., Cheng, A.H.-D. and Ouazar, D., "Analytical stochastic solutions of saltwater/freshwater interface in coastal aquifers," *Stochastic Hydrology & Hydraulics*, **12**, 413–430, 1998.

Naji, A., Cheng, A.H.-D. and Ouazar, D., "BEM solution of stochastic seawater intrusion," *Eng. Analy. Boundary Elements*, **23**, 529–537, 1999.

Nishikawa, T., "Water-resources optimization model for Santa Barbara, California," *J. Water Resour. Planning Management, ASCE*, **124**, 252–263, 1998.

Ouazar, D. and Cheng, A.H.-D., "Application of genetic algorithms in water resources," Chap. 7, In: *Groundwater Pollution Control*, ed. K.L. Katsifarakis, 293–316, WIT Press, 1999.

Ritzel, B.J., Eherat, J.W. and Ranjithan, S., "Using genetic algorithms to solve a multiple-objective groundwater pollution containment-problem," *Water Resour. Res.*, **30**, 1589–1603, 1994.

Rogers, L.L. and Fowla, F.U., "Optimization of groundwater remediation using artificial neural networks with parallel solute transport modeling," *Water Resour. Res.*, **30**, 457–481, 1994.

Sawyer, C.S. and Lin Y., "Mixed-integer chance-constrained models for groundwater remediation," *J. Water Resour. Planning Management, ASCE,* **124**, 285–294, 1998.

Shamir, U., Bear, J. and Gamliel, A., "Optimal annual operation of a coastal aquifer," *Water Resour. Res.,* **20**, 435–444, 1984.

Sorek, S. and Pinder, G.F., "Survey of computer codes and case histories," Chap. 12, In: *Seawater Intrusion in Coastal Aquifers—Concepts, Methods, and Practices,* eds. J. Bear, A.H.-D. Cheng, S. Sorek, D. Ouazar and I. Herrera, Kluwer, 403–465, 1999.

Strack, O.D.L., "A single-potential solution for regional interface problems in coastal aquifers", *Water Resour. Res.,* **12**, 1165–1174, 1976.

Tiedeman, C. and Gorelick, S.M., "Analysis of uncertainty in optimal groundwater contaminant capture design," *Water Resour. Res.,* **29**, 2139–2153, 1993.

Tung, Y., "Groundwater management by chance-constrained model," *J. Water Resour. Planning Management, ASCE,* **112**, 1–19, 1986.

Wagner, B.J., "Evaluating data worth for groundwater management under uncertainty," *J. Water Resour. Planning Management, ASCE,* **125**, 281–288, 1999.

Wagner, B.J. and Gorelick, S.M., "Optimal groundwater quality management under parameter uncertainty," *Water Resour. Res.,* **23**, 1162–1174, 1987.

Willis, R. and Finney, B.A., "Planning model for optimal control of saltwater intrusion," *J. Water Resour. Planning Management, ASCE,* **114**, 333–347, 1988.

Willis, R. and Yeh, W.W.-G., *Groundwater Systems Planning and Management,* Prentice-Hall, 1987.

Young, R.A. and Bredehoe, J.D., "Digital-computer simulation for solving management problems of conjunctive groundwater and surface water systems," *Water Resour. Res.,* **8**, 533–556, 1972.

CHAPTER 12

Hydrogeological Investigations and Numerical Simulation of Groundwater Flow in the Karstic Aquifer of Northwestern Yucatan, Mexico

L.E. Marin, E.C. Perry, H.I. Essaid, B. Steinich

1. INTRODUCTION

   The aquifer in northwestern Yucatan contains a freshwater lens that floats above a denser saline water wedge that penetrates more than 40 km inland [Back and Hanshaw, 1970; Durazo *et al.*, 1980; Back and Lesser, 1981; Gaona *et al.*, 1985; Perry *et al.*, 1989]. Recently, it has been shown that the penetration is more than 110 km [Perry *et al.*, 1995; Steinich and Marin, 1996]. The aquifer, which is unconfined except for a narrow band along the coast [Perry *et al.*, 1989], is the sole freshwater source in northwestern Yucatan. Development of industry and agriculture, and other land use changes, pose a potential threat to the quantity and quality of freshwater resources in the Yucatan Peninsula. This chapter reports field investigations used for the construction of a groundwater flow model developed for the purpose of increasing our understanding of the groundwater system, and estimating the hydraulic response to aquifer stresses. The groundwater flow model is also useful in detailed studies of saltwater intrusion, and the tracking of contaminants from industrial or agricultural sources. Ultimately, it can serve as a basic information source for local groundwater resources management.

   The objectives of this research are to: (1) describe the hydrogeologic system for northwestern Yucatan including the identification of hydrogeologic boundaries; (2) determine whether it is possible to simulate groundwater flow using a sharp interface model in this karstic aquifer; and (3) examine how the system responds to stresses such as breaching of the coastal aquitard.

1-56670-605-X/04/$0.00+$1.50
© 2004 by CRC Press LLC

2. PREVIOUS STUDIES

The hydrogeology of the eastern coast of the Yucatan Peninsula has been extensively studied by Back and Hanshaw [1970], Weidie [1982], Back et al., [1986], Stoessell et al., [1990], and Moore et al., [1992]. The hydrogeology of the northwestern part of the Yucatan Peninsula, however, has received little attention until recently [Perry et al., 1989, 1990; Marin, 1990; Marin et al., 1990; Steinich and Marin, 1996, 1997]. Back and Hanshaw [1970] called attention to important characteristics of the hydrogeology of Yucatan such as the high permeabilities found in this area and the presence of a saltwater wedge that extends tens of kilometers inland. They observed that no integrated drainage system existed in northwestern Yucatan, and that no rivers existed in this part of the peninsula. They also inferred a low gradient of the water table (based on the very low topographic relief), a high permeability of the aquifer, which they suggested probably contained large interconnected openings. Assuming that no confining beds were present (due to the thin freshwater lens), they suggested that groundwater flowed in a north-northeastern direction. The upper geologic section of the northern Yucatan Peninsula consists of nearly flat-lying carbonate, evaporitic rocks, and sediments [Lopez Ramos, 1973].

Stoessell et al. [1990] discussed hydrogeochemical and hydrogeologic features of the east coast of the Yucatan Peninsula, which differed significantly in its hydrogeologic characteristics from the north coast. Aspects particular to the hydrogeology of the northwestern Yucatan coast have been described by Perry et al. [1989, 1990, 1995] and Steinich and Marin [1996, 1997]. One of the main differences between the east coast and the north coast is that, in northwestern Yucatan, there is a narrow, chemically produced aquitard that separates the freshwater lens below from unconfined saline groundwater above. A summary of the permeability characteristics of the northwestern Yucatan Peninsula is presented in Table 1.

Chappell and Shackleton [1986] have shown that sea level oscillated at approximately 50 m below present mean sea level (MSL) between 35,000 and 120,000 years before the present. This suggests that considerable secondary cavern porosity and permeability may have developed (in a zone below present sea level) during this late Pleistocene period of stasis. It further suggests that there may exist a layer of high permeability at depth. There is limited evidence of a high permeability layer 50 m below MSL [Gmitro, 1987; Rosado, 1987; Marin, 1994].

Location	Mérida Block	Ring of Cenotes	North Coast Confining Layer*
References	[Marin *et al.*, 1990]	[Marin *et al.*, 1990]	[Perry *et al.*, 1989, 1990; Marin *et al.*, 1988]
Geologic/ Hydrogeologic Features	Intergranular permeability dominant. Block consists of highly permeable sedimentary rocks.	High cavern permeability inferred from abundance of cenotes and caves.	Near-surface aquitard that divides saltwater (above) from fresh/brackish water (below). (Both water layers overly saltwater intrusion.)
Physiographic Examples/ Evidence	Flat, immature karst surface, relatively few cenotes or caves.	Many cenotes aligned in a semicircle of radius 90 km.	Petenes (flowing springs that are cenotes drowned by rising sea level/rising water table).
Hydrogeologic Characteristics	Flat water table (typical gradient 7–10 mm/km). Water table responds quickly and uniformly to seasonal or local precipitation.	High groundwater flow; abundant springs where Ring intersects coast.	Confined water transmits tidal pressure for up to 20 km inland.

* Overlies part of Ring of Cenotes and Mérida Block

Table 1: Hydrogeologic characteristics of the Yucatan Peninsula.

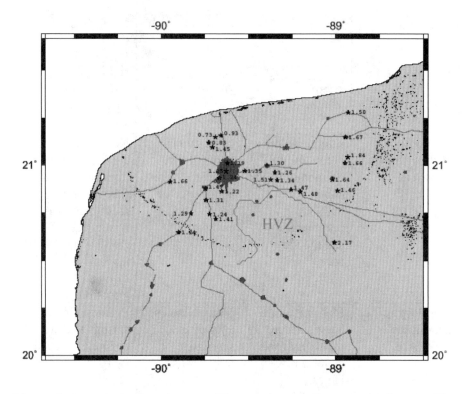

Figure 1: Location of study area. The continuous lines are highways. The shaded region delineates the approximate location of the Ring of Cenotes. (Also shown is the "Highly Variable Zone" discussed in the text.)

3. HYDROGEOLOGIC STUDIES

3.1   Hydrogeologic Setting

We propose that the northwestern Yucatan Peninsula contains three somewhat overlapping zones (Figure 1), differing by the type of permeability (Table 1). A large and hydrogeologically homogeneous part of the northwest Peninsula, here labeled "Mérida Block", lies within a semicircle of approximately 180 km diameter centered at about 35 km north-northeast of Mérida. This is bounded by the second zone, which has become known as the "Ring of Cenotes" (cenote = sinkhole), a 5–20-km wide band (Figure 1 [Marin et al., 1990]). The hydrogeologic properties and their significance are described in the next section. The third zone is the north coast-confining layer, which is distinguished by a near-surface aquitard that affects both the piezometric head, and the thickness of the coastal edge of the freshwater lens.

The north coast confining layer is a unique, chemically produced layer that forms a band several km wide along much of the north Yucatan coast from Celestun to the east of Dzilam Bravo (Figure 1) [Perry et al., 1989; Tulaczyk et al., 1993; Smart and Whitaker, 1990; Perry et al., 1990]. Perry et al. [1989] postulated that the 0.5 m thick confining layer, found at depths that range from the surface to 5 m below, has been produced behind the north coast dune in a zone (tsekel) where the freshwater table intersects and moves seasonally across the gently sloping (approximately 20 cm/km) land surface. Here, $CaCO_3$-saturated groundwater precipitates calcite in small pore spaces of exposed rock (but not in large cavities such as the drowned cenotes that form springs (petenes) [Marin et al., 1988]). The result of this precipitation is a thin, nearly impermeable calcrete aquitard. Presumably, this layer has propagated inland during the last 5000–6000 years of slowly rising sea level [Coke et al., 1990]. The coastal confining layer causes a thickening of the freshwater lens [Perry et al., 1989; Marin, 1990; Tulaczyk et al., 1993] so that in the north coast fishing port of Chuburna (for example), just west of Progreso (Figure 1), the lens has a calculated thickness of about 18 m at the shore.

A first-order topographic survey of most of the northwest study area [Echeverria, 1985; Echeverria and Cantun, 1988] makes possible the determination of the extremely flat hydraulic gradients (on the order of 5–10 mm/km [Marin et al., 1987; Marin, 1990]) of the area. The low gradient, which is difficult to measure, suggests very high permeabilities. Sampling points were the shallow private wells present in many towns and cities. These wells typically are hand-dug, have an approximate diameter of 1 m, and are finished 0.5–1.0 m below the water table.

From this survey, Marin [1990] established water-level elevations for a network of more than 100 points. Water levels at these stations were measured one to six times (July, 1987; January, April, July, and September, 1988; April, 1989); and water table maps of northwestern Yucatan have been prepared for those dates. Figure 2 shows the water table for July 1987. This map was chosen because it is representative of the water table in Yucatan for the study period. Measured heads in northwestern Yucatan range from a low of 0.45 m above MSL near Chuburna to a high of 2.1 m above MSL in Sotuta on the southeastern portion of the study area. Depth to the water table ranges from the surface along the coast to 18 m at Sotuta (Figure 1) 60 km inland. During the period of observation, variations in the water table between the dry and wet seasons ranged from 5 to 61 cm during the study period (which was less than 2 years) that water levels were measured. Steinich and Marin [1997] have identified an area in the aquifer where there are important variations in the water levels within a short period of time.

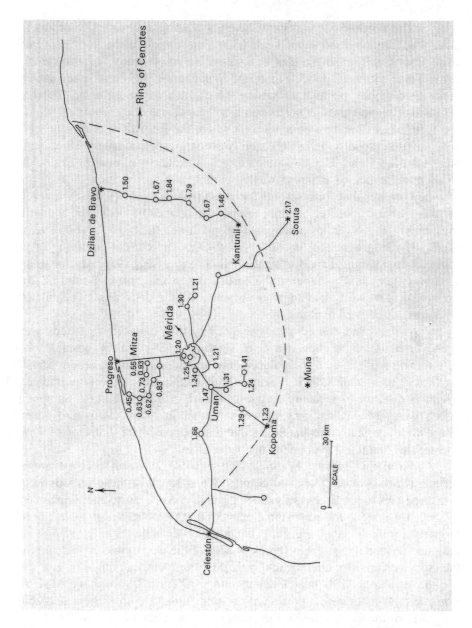

Figure 2: Water table map for northwest Yucatan. Note the low elevation of the water table above MSL and the very low hydraulic gradient (average 10 mm/km, over the region). (Reprinted with permission.)

They have identified this zone as the "Highly Variable Zone" (Figure 1). Water levels on the eastern side of the study area are higher than those in the central region (Figure 2). This is probably a reflection of the spatial distribution of precipitation on the Yucatan Peninsula. The average annual precipitation along the eastern coast of the peninsula is on the order of 1,500 mm, whereas the average annual precipitation at Progreso (Figure 1) is 500 mm [INEGI, 1981]. Evapotranspiration has been reported to be 90–95% of the precipitation that falls on the Yucatan Peninsula [INEGI, 1983].

3.2   Hydrogeologic Boundaries

Two hydrogeologic boundaries were identified: the Ring of Cenotes and the Gulf of Mexico. The alignment of cenotes appears in the geologic map published by the Instituto Nacional de Estadistica, Geografia, e Informatica [INEGI, 1983]. The Ring of Cenotes, (hereafter "Ring"), which is a remarkably regular circular arc, has recently been attributed to enhanced permeability associated with a large extraterrestrial impact structure formed at the end of the Cretaceous Period [Pope *et al.*, 1991; Perry *et al.*, 1995; Hildebrand *et al.*, 1991; Sharpton *et al.*, 1992, 1993]. The Ring is located between the second and the third ring of the Chicxulub Multiring Impact Basin as defined by Sharpton *et al.* [1993]. The association of the Ring with the buried impact structure bears on the regional hydrogeology because it implies that the high permeability of the Ring is ultimately controlled by relatively deep subsurface geologic features that are not subject to direct observation [Perry *et al.*, 1995; Steinich and Marin, 1996]. The hypothesis of deep control over permeability is supported by the observation that at least one cenote of the Ring (Xcolak, Figure 1) extends vertically for 120 m below the present water table. Presumably, such a vertical shaft could only develop within the vadose zone where downward movement of water prevails [Noel and Choquette, 1987]. This implies an extensive, deep zone of high permeability associated with a paleo-water table much lower than the present water table.

The Ring is a zone of high permeability as shown by: (1) transects characterized by a decline in water levels toward the Ring (Figures 3a and b) and (2) high density of springs and breaks on sand bars at the intersection of the Ring with the sea. Thus, the Ring affects groundwater flow by diverting some or all of the groundwater flowing across the Ring and discharging it to the sea [Marin, 1990; Marin *et al.*, 1987, 1990]. Evidence supporting this hypothesis also comes from Perry *et al.* [1995] and from Velazquez [1995], who found a similar Cl^-/SO_4^{2-} ratio in the Ring near Kopoma as well as near Celestun, and also from Steinich and Marin [1996], who determined that the Ring south of Mérida is a high permeability zone, using electrical methods.

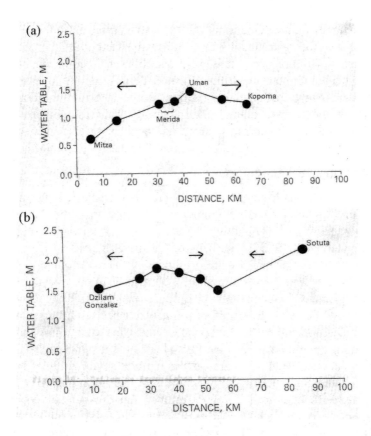

Figure 3: (a) Mitza-Kopoma and (b) Dzilam Gonzalez-Sotuta transect. (Water levels increase with distance away from the sea. Water levels decrease as the Ring is intersected and continue to increase with distance away from the sea. Arrows indicate groundwater flow directions.) (Part (a) from Steinich and Marin (1996), with permission.)

Since little question remains that the Ring of Cenotes is related to the buried Chicxulub Impact Structure, it can be presumed that the high permeability zone extends hundreds of meters into the subsurface. This observation is corroborated with the geochemical and geoelectrical data [Perry *et al.*, 1995; Velázquez, 1995; Steinich and Marin, 1996]. The origin of this Ring is discussed elsewhere [Pope *et al.*, 1991; Perry *et al.*, 1995].

The Gulf of Mexico forms a natural hydrogeologic boundary of the study area on the north and west. The Ring, which acts as a high permeability zone, affects groundwater flow to the south and east. This was established by the two north–south transects crossing the Ring (Figures 3a and b). Water levels increase with distance away from the coast for 40–60 km (San Ignacio-Kopoma transect) and for 30 km (Dzilam Gonzalez-Sotuta transect); but still farther south, water levels decrease slightly until the transects cross the Ring. A third transect, an east–west transect located on the northeastern section of the study area, shows the same behavior. These patterns were observed for almost 2 years (1987–1989). These results support the hypothesis that the Ring is a zone of high permeability with respect to its surroundings. The Ring does not, however, affect groundwater flow equally throughout the Ring. Steinich *et al.* [1996] have identified the groundwater divide within the Ring of cenotes with a study that combined hydrogeology and geochemistry. Directly south of Mérida, along the western boundary of the "Highly Variable Zone," there is a mound along the southeastern portion of the study area suggesting that water may flow into the study area near Kantunil from a bordering region of higher recharge about 55 km from the coast as well as from the groundwater divide [Marin, 1990; Steinich *et al.*, 1996; Steinich and Marin, 1997].

3.3   Geometry of Freshwater Lens

The thickness of the freshwater lens was estimated from measured water levels using the Ghyben-Herzberg relation, which balances a column of seawater with an equivalent fresh/saltwater column. This relation assumes that simple hydrodynamic conditions exist, that the boundary separating the fresh and saltwater layers is sharp, and that there is no seepage face [Freeze and Cherry, 1979]:

$$z = \frac{\rho_f}{\rho_s - \rho_f} h_f \tag{1}$$

where z = thickness of the freshwater lens from the interface to mean sea level (MSL)

ρ_f = density of freshwater, assumed to be 1.000 g/cm^3

ρ_s = density of saltwater, assumed to be 1.025 g/cm^3

h_f = freshwater head above MSL

Substituting the values for ρ_f and ρ_s one has:

$$z = 40 h_f \tag{2}$$

Thus, the depth of freshwater length to the interface is 40 times the freshwater head.

Water elevation data of July 1987 was used to calculate the thickness of the freshwater lens. July measurements were chosen because it is about the middle of the May-through-September rainy season; thus it is about midway through the annual recharge cycle. The postulated geometry of

Location	Date	Head (m) above MSL	Depth to interface below MSL (m)	
			measured	calculated
Mérida*	4/89	0.96	37	38
Noc-Ac	4/89	0.84	>27	34
Dzibilchaltun	7/89	0.73	>27	28
MITZA	7/88	0.55	>15	22
Labon	7/89	1.58	>40, <50	62

* Depth to interface measured by Villasuso (personal communication).

Table 2: Measured interface depths vs. those calculated using the Ghyben-Herzberg principle. (Note: the top of interface was located at 27 m at cenote Noc-Ac. The interface was not reached at Dzibilchaltun. MITZA is a man-made lake.)

the freshwater lens is shown in Figure 3. Note that the thickness of the lens should vary from a low of 18 m near Chuburna along the coast to more than 80 m in Sotuta, located in the southeastern portion of the study area. Limited data (Table 2) suggest that the Ghyben-Herzberg relation does not significantly overestimate the thickness of the freshwater lens in northwestern Yucatan. Recent work [Steinich and Marin, 1996] in which electrical resistivity surveys were correlated with water level measurements have shown that the Ghyben-Herzberg relation holds well for northwestern Yucatan.

3.4 Conceptual Model

Following is a description of the conceptual model used to simulate groundwater flow in northwestern Yucatan. The aquifer is unconfined except for a narrow band parallel to the coast. This confining layer extends on the order of 5 km seaward. Recharge occurs throughout the aquifer, with water flowing from south to north, except for a zone parallel to the Ring of Cenotes, where the groundwater flow direction is reversed (Figures 2 and 3). Discharge from the aquifer occurs throughout the coast, with a higher concentration occurring at the two intersections of the Ring of Cenotes with the sea. The aquifer was assumed to be heterogeneous, with the Ring of Cenotes being a higher permeability zone (one order of magnitude higher than the surrounding area). The aquifer was assumed to behave as an equivalent porous media. The aquifer was simulated using a two-layer model with a layer of high permeability 50 m below the present surface. This assumption is justified since the sea level has oscillated at this depth between the last 35,000 and 120,000 years. Recharge varied from 100 to 220 mm/yr.

It follows the same spatial distribution as the precipitation, according to the INEGI [Anonymous, 1980] maps.

4. NUMERICAL MODELING

The numerical model used for the Yucatan aquifer was SHARP, a quasi-three-dimensional finite difference model for the simulation of fresh- and saltwater flow in a coastal aquifer system [Essaid, 1990]. Large Representative Elementary Volumes (REVs) were used to treat the simulated area as an equivalent porous media [Marin, 1990]. Gonzalez-Herrera [1992], who has subsequently attempted to model groundwater flow in this karstic aquifer, has also approached the problem using large REVs. The model SHARP is quasi-three-dimensional because it assumes horizontal flow in the aquifers and vertical flow in the confining layers. The model uses two governing equations, one for the freshwater domain and one for the saltwater domain. The fresh- and saltwater flow equations, coupled at the interface, are integrated over the vertical dimension because it is assumed that there are no vertical gradients within the aquifer (Dupuit assumption). The model may be used for a heterogeneous, anisotropic, multi-aquifer system. The governing equations are [Essaid, 1990]:

$$S_f B_f \frac{\partial h_f}{\partial t} + n\alpha \frac{\partial h_f}{\partial t} + \left[n\delta \frac{\partial h_f}{\partial t} - n(1+\delta) \frac{\partial h_s}{\partial t} \right]$$
$$= \frac{\partial}{\partial x} \left(B_f K_{fx} \frac{\partial h_f}{\partial x} \right) + \frac{\partial}{\partial y} \left(B_f K_{fy} \frac{\partial h_f}{\partial y} \right) + Q_f + Q_{lf} \tag{3}$$

$$S_s B_s \frac{\partial h_s}{\partial t} + \left[n(1+\delta) \frac{\partial h_s}{\partial t} - n \frac{\partial h_f}{\partial t} \right]$$
$$= \frac{\partial}{\partial x} \left(B_s K_{sx} \frac{\partial h_s}{\partial x} \right) + \frac{\partial}{\partial y} \left(B_s K_{sy} \frac{\partial h_s}{\partial y} \right) + Q_s + Q_{ls} \tag{4}$$

where h_f is the freshwater head, h_s is the saltwater head, S_f is the freshwater specific storage, S_s is the saltwater specific storage, B_f is the thickness of the freshwater zone, B_s is the thickness of the saltwater zone, t is time, $\delta = \rho_f / (\rho_s - \rho_f)$, K_{fx} and K_{sx} are the fresh- and saltwater hydraulic conductivities in the x-direction (LT^{-1}), K_{fy} and K_{sy} are those in

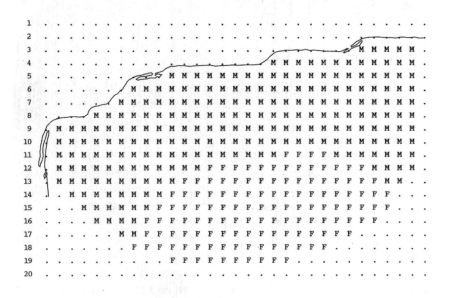

Figure 4: Study area with model grid. M denotes the presence of mixed waters (freshwater and saltwater); F denotes the presence of freshwater only. The first active node along the top is located offshore; the active nodes located on the eastern, southern, and western boundaries correspond to the "Ring of Cenotes."

the y-direction, Q_f and Q_s are the fresh- and saltwater source/sink terms (LT^{-1}), Q_{lf} and Q_{ls} are the fresh- and saltwater leakage terms (LT^{-1}), the parameter α is given the value of 1 for an unconfined aquifer and 0 for confined, and n is the porosity.

4.1   Model Framework

The study area was divided into a grid of 19 rows by 31 columns (Figure 4). The horizontal cell dimensions were 6.3 by 6.3 km. The model SHARP isolates the study area by imposing no-flow boundary conditions around it. Because the boundaries were considered remote to the particular area of interest, the city of Mérida and the north coast, the condition was justified except for the northern boundary. The boundary condition to the north was a head-dependent boundary (Gulf of Mexico). For the head-dependent boundary, equivalent freshwater heads were specified for the offshore nodal areas. This allowed for the leakage of freshwater through the coastal aquitard to the sea.

Parameter	Value	Reference
Hydraulic Conductivity (m/s)	$3\times10^{-4}-5\times10^{-1}$	Reeve and Perry [1990]
	10^{-2}	Freeze and Cherry [1979]
	10^{-2}	Back and Lesser [1981]
	$10^{-6}-5\times10^{-3}$	Gonzalez-Herrera [1984]
Porosity (%)	7–41	Gonzalez-Herrera [1984]
Recharge (mm/yr)	100–200	Anonymous [1980]

Table 3: Literature values for aquifer parameters.

For the purpose of these simulations, we assumed that: (a) the aquifer is heterogeneous (with a higher permeability along the Ring); (b) the aquifer is isotropic within each layer; (c) the aquifer has a sharp interface dividing the fresh- and saltwater; (d) the aquifer is unconfined except near the coast; and (e) the coastal confining layer described by Perry *et al.* [1989] starts 6.3 km from the coast and extends 6.3 km seaward (i.e., one node offshore, and the first inland node).

4.2   Model Calibration and Sensitivity

Aquifer parameters selected from the literature were initially used for calibration of the model. These are given in Table 3. The range in hydraulic conductivity given by Reeve and Perry [1990] was determined for the aquifer near Chuburna, just north of Mérida, 18 km west of Progreso. The value from Freeze and Cherry [1979] is their reported high value given for karstic terrains. The value from Back and Lesser [1981] was back-calculated from the annual average discharge they reported per kilometer of coast. The permeabilities listed by Gonzalez-Herrera [1984] are for laboratory cores taken from wells in Mérida that ranged from 10 to 80 m deep. The laboratory core measurements were minimum values because they did not include the permeability associated with fractures, conduits, and caverns.

Model sensitivity analysis consisted of a determination of the effects of varying the following parameters with respect to the simulated water table elevations: depth of high permeability zone, recharge, hydraulic conductivities, and use of one- versus two-layer model [Marin, 1990]. Better results were obtained using a two-layer model. As part of the sensitivity analysis, it was determined that the two most important parameters were the distribution of the hydraulic conductivity and the recharge.

Best results were obtained using a two-layer model with a high permeability layer overlain by a layer of lower permeability. The thickness

Parameter	Layer 2 (lower)	Layer 2 (upper)
Hydraulic Conductivity	1 m/s	0.1 m/s
Thickness	150 m	55 m
Recharge		100–300 mm/yr

Table 4: Parameters used for steady-state simulation of the aquifer in northwest Yucatan.

selected for the bottom layer was 300 m and that of the upper layer was 50 m. Table 4 shows the parameters used for the steady-state solution. The time needed to achieve steady state under these conditions was 25 years. Geologically, the two-layer model may be justified based on past sea level stands. We propose that a high permeability layer developed at approximately 50 m below MSL as a result of chemical erosion taking place at the paleo water table.

4.3   Discussion of Numerical Simulation

The predicted head distribution from the two-layer model compares favorably to the field data (Figure 6). The hydraulic conductivity for both the x (E–W) and y (N–S) directions of the lower layer was 1 m/s and that of the upper layer 0.1 m/s (for both x and y directions) except for a 6.3 km band representing the Ring that was assigned a hydraulic conductivity of 1 m/s.

The steady-state simulation of groundwater flow in northwest Yucatan predicts a distribution of the saltwater intrusion that is consistent with field data (Figure 4) [Back and Hanshaw, 1970; Perry et al., 1989; Steinich and Marin, 1996].

4.4   Simulation of Breached Confining Layer

The Mexican government is committed to developing tourist complexes and to build shelter ports along the coast of Yucatan. In order to do this, the confining layer described by Perry et al. [1989] will be breached, since the sites where the shelter ports are built must be excavated; thus, the confining layer will be destroyed. Perry et al. [1989] postulated that continued breaching of the confining layer would result in a partial collapse of the freshwater lens. This hypothesis was tested by comparing the results of a simulation with the confining layer and without it.

The breaching of the confining layer was simulated by setting a high leakance value. When a high leakance value is specified, the value in the offshore nodal area defaults to the constant head value given in the input file. In Figure 5, once the confining layer has been breached, the coastal nodal areas default to 0.25 m. The drop of head ranges from approximately 25 cm

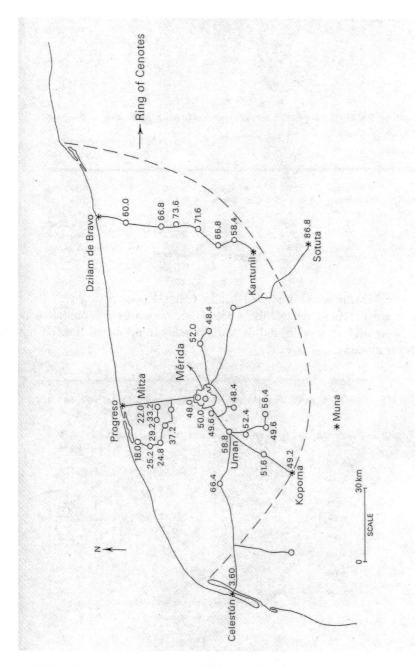

Figure 5: Estimate of the depth of the fresh/saltwater interface below MSL. (Water levels from July 1987 were used to estimate freshwater lens. Lens in northwestern Yucatan is less than 100 m throughout studied area.)

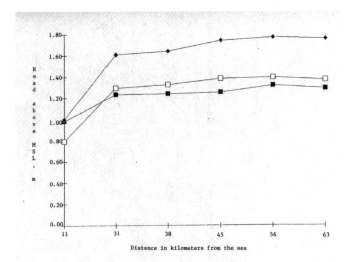

Figure 6: Measured vs. predicted heads: Mitza-Kopoma transect. The dark squares are the field data (July, 1988). Open squares are predicted heads from the two layer model, and dark diamonds are predicted heads from the one-layer model.

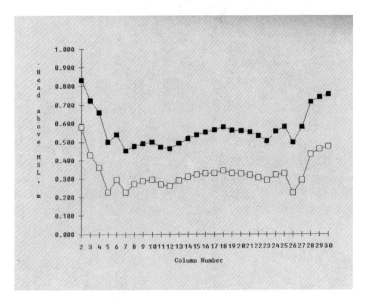

Figure 7: Predicted heads: confined (dark squares) vs. breached (open squares) heads at coastal nodes. Notice increased discharge at the intersection of the "Ring of Cenotes" with the sea (at either end of the graph).

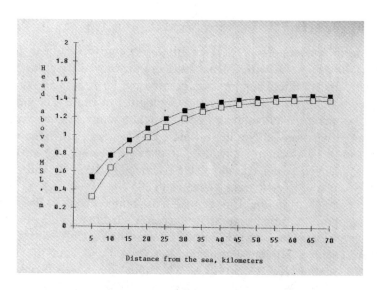

Figure 8: Cross section showing effect of breaching layer. (Maximum effect occurs at the coast (drop of 30 cm in head).)

to 55 cm. If these predictions are accurate, this would result in a loss of approximately 12 m of freshwater at the coast. A head loss of 30 cm at Chuburna would result in a freshwater lens of only 10 m as opposed to the estimated 20 m at the present [Perry *et al.*, 1989]. According to this model, the thickness of the whole lens would decrease, with the most dramatic impact found along the coast (Figures 7 and 8).

5. CONCLUSIONS

The water table maps for northwestern Yucatan reveal a very low hydraulic gradient, indicating very high permeabilities. The calculated thickness of the freshwater lens using the Ghyben-Herzberg ratio varies between a low of 18 m near the coast to over 80 m more 60 km inland. The large REVs used for this study justified the use of a "porous media" approach. Both the field and the simulated data were consistent. Thus, the model was used in a predictive mode. This model supports the hypothesis that continued breaching of the confining layer may result in a loss of head of up to 30 cm in the coast. This loss would correspond to a freshwater loss of about 12 m. The Mexican National Commission on Water (CNA) has been informed about this model and the predictions made with it.

Acknowledgments

Marin was supported with an Illinois Minority Graduate Incentive Program fellowship and with a Doctoral Completion Award from Northern Illinois University. Field support was granted by the American Association of Petroleum Geologists, The American Geological Institute, The Geological Society of America, The National Speleological Society, Sigma Xi, and the Department of Geology at Northern Illinois University. Marin acknowledges support from the Dirección General de Asuntos del Personal Académico of the Universidad Nacional Autónoma de México (IN106891) and from the Consejo Nacional de Tecnología y Ciencia (T2057).

Personnel from the Universidad Autónoma de Yucatan provided logistical support. Our thanks to M. Villasuso, J. Gamboa, and V. Coronado. Perry acknowledges support from the Northern Illinois University Graduate School Research Fund, the Petroleum Research Fund, and the National Science Foundation (EAR 8508173). The authors wish to thank Steven Philips and Ward Sanford for reviewing a previous version of this manuscript.

REFERENCES

Anonymous, unpublished report on file, INEGI, Paseo Montejo, Mérida, Yucatán, México, 1980.

Back, W. and B. Hanshaw, "Comparison of chemical hydrogeology of the carbonate peninsulas of Florida and Yucatan," *J. Hydrology*, **10**, 330–368, 1970,

Back, W. and J.M. Lesser, "Chemical constraints of groundwater management in the Yucatan Peninsula, Mexico," In: L.R. Beard, ed., Water for Survival, *J. Hydrology*, **51**, 119–130, 1981.

Back, W., B. Hanshaw, J.S. Herman, and J.N. Van Driel, "Differential dissolution of a Pleistocene reef in the groundwater mixing zone of coastal Yucatan, Mexico," *Geology*, **14**, 137–140, 1986.

Chappell, J. and N.J. Shackleton, "Oxygen isotopes and sea level," *Nature*, **324**, 137–140, 1986.

Coke, J. E.C. Perry, and A. Long, "Charcoal from a probable fire pit in the Yucatan Peninsula, Mexico: another point in the glacio-eustatic sea level curve," *Nature*, **353**, p. 25, 1990.

Durazo, J., S. Gaona, J. Trejo, and M. Villasuso, "Observaciones sobre la interfase salina en dos cenotes del centro-norte de Yucatán," XXIII Congreso Nacional de Investigación en Física: Sociedad Mexicana de Física, Guadalajara, Jalisco, 24–28 November, 1980.

Echeverria, E, Unpublished report on file, INEGI, Paseo Montejo, Mérida, Yucatán, México, 1985.

Echeverria, E. and C. Cantun, Unpublished report on file, INEGI, Paseo Montejo, Mérida, Yucatán, México, 1988.

Essaid, H.I, "The computer model, SHARP, a quasi-3-D finite-difference model to simulate freshwater and saltwater flow in layered coastal aquifer systems," U.S.G.S. Water Resources Investigations Report 90-4130, 181 p., 1990.

Freeze, R.A., and J.A. Cherry, *Groundwater*, Prentice-Hall, 604 p., 1979.

Gaona, S., M. Villasuso, J. Pacheco, A. Cabrera, J. Trejo, T. Gordillo de Anda, C. Tamayo, V. Coronado, J. Durazo, y E.C. Perry, Hidrogeoquímica de Yucatán 1: Perfiles hidrogeoquímicos rofundos en algunos lugares del acuífero del noroeste de la Península de Yucatán, Boletín, Instituto de Geofísica-UNAM, México, 1985.

Gmitro, D.A, "The interaction with carbonate rock in Yucatan, Mexico," M.S. thesis, Northern Illinois University, DeKalb, IL, 111 p., 1987.

Gonzalez-Herrera, R.A, "Correlación de muestras de roca en pozos de la Ciudad de Mérida," B.S. thesis, Universidad Autónoma de Yucatán, Mérida, Yucatán, México, 1984.

Gonzalez-Herrera, R.A, "Evolution of groundwater contamination in the Yucatan Karstic aquifer," M.S. thesis, University of Waterloo, Waterloo, Canada, 146 p., 1992.

Hildebrand, A. R., G.T. Penfield, D.A. Kring, M. Pilkington, A. Camargo Z., S.B. Jacobsen, and W.V. Boynton, *Geology*, **19**, 867–871, 1991

INEGI, "Carta de precipitación de Yucatán, 1:1,000,000," Instituto Nacional de Estadística, Geografía e Informática, Paseo Montejo, Mérida, Yucatán, México, 1981.

INEGI, "Carta geológica de Mérida, 1:250,000," Instituto Nacional de Estadística, Geografía e Informática, Paseo Montejo, Mérida, Yucatán, México, 1983.

INEGI, "Carta de evapotranspiración de Yucatán, 1:1,000,000," Instituto Nacional de Estadística, Geografía e Informática, Paseo Montejo, Mérida, Yucatán, México, 1983.

Lopez Ramos, E, *Geologia de Mexico*, v. 3, Tesis Resendiz, Mexico City, Mexico, 453 p., 1973.

Marin, L.E, "Field investigations and numerical simulation of the karstic aquifer of northwest Yucatan, Mexico," Ph.D. thesis, Northern Illinois University, DeKalb, IL, 183 p., 1990.

Marin, L.E, "The hydrostratigraphy of Yucatan," Abstract, Asociación Mexicana de Hidráulica—sección Sureste, Mérida, Yucatán, May 6, 1994.

Marin, L.E., E.C. Perry, C. Booth and M. Villasuso, "Hydrogeology of the northwestern Peninsula of Yucatan, Mexico," *EOS*, Transactions, American Geophysical Union, **69**(16), 1292, 1987.

Marin, L.E., R. Sanborn, A. Reeve, T. Felger, J. Gamboa, E.C. Perry, and M. Villasuso, "Petenes: a key to understanding the hydrogeology of Yucatán, Mexico," Abstract, International Symposium on the Hydrogeology of Wetlands in Semi-Arid and Arid Areas, Seville, Spain, May 9–12, 1988.

Marin, L.E.; E.C. Perry, K.O. Pope, C.E. Duller, C.J. Booth, and M. Villasuso, "Hurricane Gilbert: its effects on the aquifer in northern Yucatan, Mexico," In: Selected papers on hydrogeology from the 28th International Geological Congress, Washington, D.C., E.S. Simpson and J.M. Sharp, Jr. eds. Verlag Heinz Heise, Hannover. p. 111–128, 1990.

Moore, Y.H., R.K. Stoessell, and D.H. Easley, "Freshwater/sea-water relationship within a ground-water flow system, northeastern coast of the Yucatan Peninsula," *Ground Water*, **30**, 343–350, 1992.

Noel, P.J. and P.W. Choquette, Diagenesis "9—limestones—the meteoric diagenetic environment," *Canadian Geoscience*, **11**, 161–194, 1987.

Perry, E.C., J. Swift, J. Gamboa, A. Reeve, R. Sanbonr, L.E. Marin, and M. Villasuso, "Environmental aspects of surface cementation, north coast, Yucatan, Mexico," *Geology*, **17**, 818–821, 1989.

Perry, E.C., A. Reeve, L.E. Marin, and M. Villasuso, "Reply to Comment on 'Environmental aspects of surface cementation, north coast, Yucatan, Mexico'", *Geology*, September, 1990.

Perry, E.C., L.E. Marin, J. MacLain, and G. Velazquez, "The influence of the Chicxulub crater on regional hydrogeology of northwest Yucatan, Mexico," *Geology*, 1995.

Pope, K.O., A. C. Ocampo, and C. E. Duller, "Mexican site for K/T impact crater?" *Nature*, **351**, 105, 1991.

Reeve, A. and E.C. Perry, "Aspects and tidal analysis along the western north coast of the Yucatan Peninsula, Mexico," AWRA: International Symposium on Tropical Hydrogeology and Fourth Caribbean Islands Water Resources Conference, San Juan, Puerto Rico, July 23–27, 1990.

Rosado, F., oral communication, 1987.

Sharpton, V.L., G. B. Dalrymple, L.E. Marin, G. Ryder, B. C. Schuraytz, and J.Urrutia-Fucugauchi, "New links between the chicxulub impact structure and the cretaceous-tertiary boundary," *Nature*, **359**, 819–821, 1992.

Sharpton, V.L., K. Burke, A. Camargo, S.A. Hall, L. E. Marin, G. Suárez, J.M. Quezada, P.D. Spudis, and J. Urrutia-Fucugauchi, "The gravity expression of the Chicxulub multiring impact basin: size, morphology, and basement characteristics," *Science*, **261**, 1564–1567, 1993.

Smart, P.L., and F.F. Whitaker, "Comment on 'Geologic and environmental aspects of surface cementation, north coast, Yucatan, Mexico'", *Geology*, 802–803, 1990.

Steinich, B., L.E. Marin, "Hydrogeological investigations in northwestern Yucatan, Mexico, using resistivity surveys," *Ground Water*, **34**(4), 640–646, 1996.

Steinich, B., G. Velázquez Oliman, L.E. Marin, and E.C. Perry, "Determination of the ground water divide in the karst aquifer of Yucatan, Mexico, combining geochemical and hydrogeological data," *Geofísica Internacional*, **35**, 153–159, 1996.

Steinich, B., L.E. Marin, "Determination of flow characteristics in the aquifer in northwest Yucatan, Mexico," *J. Hydrology*, **191**, 315–331, 1997.

Stoessel, R.K., W.C. Ward, B.H. Ford, and J.D.Schuffert, "Water chemistry and $CaCO_3$ dissolution in the saline part of an open-flow mixing zone, coastal Yucatan Peninsula, Mexico," *Geological Society of America Bulletin*, **101**(2), 159–169, 1990.

Tulaczyk, S., E.C. Perry, C. Duller and M. Villasuso, "Geomorphology and hydrogeology of the Holbox area, northeastern Yucatan, Mexico, interpreted from two LANDSAT[TM] images," In: Applied Karst Geology, Proceedings of the 4th Multidisciplinary Conference on Sinkholes and the Engineering and Environmental Impacts of Karst, B.F. Beck, editor, 181–188, 1993.

Velazquez, G, "Estudio geoquímico del Anillo de Cenotes, Yucatán, (Geochemical study of the Cenote Ring, Yucatan)," M.S. thesis (in Spanish), Instituto de Geofísica, Universidad Nacional Autónoma de México, Mexico City, Mexico, 78 p., 1995.

Waller, J. and J. Coke, oral and written communication, 1993.

Weidie, A.E, "Lineaments of the Yucatan Peninsula and fractures of the central Quintana Roo coast," Road Log and Supplement to 1978 Guidebook, 1982 Geological Society of America Meeting Field Trip #10-Yucatan, p. 21–25, 1982.

Index